Santosh Jaju
Sachin Untawale
Nilesh Nirwan

Dental Implant: Review of Material Properties of Resins

Santosh Jaju
Sachin Untawale
Nilesh Nirwan

Dental Implant: Review of Material Properties of Resins

LAP LAMBERT Academic Publishing

Imprint
Any brand names and product names mentioned in this book are subject to trademark, brand or patent protection and are trademarks or registered trademarks of their respective holders. The use of brand names, product names, common names, trade names, product descriptions etc. even without a particular marking in this work is in no way to be construed to mean that such names may be regarded as unrestricted in respect of trademark and brand protection legislation and could thus be used by anyone.

Cover image: www.ingimage.com

Publisher:
LAP LAMBERT Academic Publishing
is a trademark of
Dodo Books Indian Ocean Ltd. and OmniScriptum S.R.L publishing group

120 High Road, East Finchley, London, N2 9ED, United Kingdom
Str. Armeneasca 28/1, office 1, Chisinau MD-2012, Republic of Moldova, Europe
Managing Directors: Ieva Konstantinova, Victoria Ursu
info@omniscriptum.com

Printed at: see last page
ISBN: 978-620-8-45762-4

INDEX

Content **Page No.**

1. INTRODUCTION

The field of dental implants has undergone significant advancements over the past few decades, driven by the growing demand for effective and aesthetically pleasing solutions to tooth loss and oral rehabilitation. Dental implants are artificial roots, typically made of titanium, that provide a foundation for fixed or removable replacement teeth. While the use of titanium has been predominant due to its biocompatibility and mechanical properties, recent innovations have focused on enhancing the material properties of dental resins used in conjunction with these implants.

Dental resins play a crucial role in the success of dental implants, as they are used in the fabrication of implant-supported prosthetics, such as crowns, bridges, and dentures. The development of these resins aims to improve their mechanical strength, wear resistance, aesthetic qualities, and biocompatibility. Enhanced resin materials can significantly impact the overall durability and performance of dental implants, providing patients with more reliable and long-lasting dental solutions [1]. This introduction delves into the design and development of advanced dental resins, exploring the materials science behind these innovations and their implications for dental implant technology. By examining recent research and technological advancements, this study aims to highlight the potential of new resin materials to transform dental implantology, offering improved outcomes for patients and dental professionals alike.

Dental implants have revolutionized restorative dentistry, offering durable and aesthetically pleasing solutions for tooth replacement. Central to the success of these implants is the material used, which significantly impacts the implant's performance, longevity, and compatibility with the human body [2]. Among various materials, resins have emerged as a prominent choice due to their favorable properties. However, the continuous quest for improving these properties has led to innovative design and development strategies in dental implants. This introduction explores the intersection of dental implant technology and material science, focusing on how advanced design and development processes enhance the material properties of resins. It highlights the importance of these improvements in achieving better mechanical strength, biocompatibility, and aesthetic outcomes. By delving into the advancements in resin formulations and the integration of novel technologies, we aim to provide a comprehensive overview of the current state and future directions in the field of dental implant materials [3].

1.1 Overview of Dental Implants

Dental implants represent a significant advancement in restorative dentistry, offering a durable and aesthetically pleasing solution for tooth replacement. They serve as artificial tooth roots, usually made of biocompatible materials such as titanium, which integrate with the jawbone to provide a stable base for fixed or removable replacement teeth [4]. Dental implants closely mimic the function and appearance of natural teeth, offering numerous benefits over traditional dentures and bridges.

Components of Dental Implants, Dental implants consist of three primary components:

- **Implant Fixture:** This is the screw-like component that is surgically inserted into the jawbone. It acts as the root of the artificial tooth, providing a strong foundation.
- **Abutment:** The abutment is placed on top of the implant fixture. It serves as a connector, securing the prosthetic crown to the implant fixture.
- **Prosthetic Crown:** This is the visible part of the implant that resembles a natural tooth. It is custom-made to match the color and shape of the patient's natural teeth.

Types of Dental Implants

Dental implants can be categorized based on their placement and the specific needs of the patient:

- **Endosteal Implants:** These are the most common type of dental implants, placed directly into the jawbone. They are typically made of titanium and can hold one or more artificial teeth.
- **Subperiosteal Implants:** These implants are placed under the gum but above the jawbone. They are used for patients who have insufficient bone height and cannot undergo bone augmentation procedures.

- **Zygomatic Implants:** These are longer implants anchored in the cheekbone (zygoma) rather than the jawbone, suitable for patients with severe bone loss in the upper jaw.

Benefits of Dental Implants, Dental implants offer several advantages over other tooth replacement options:

- **Improved Appearance:** Dental implants are designed to look and feel like natural teeth, enhancing the patient's smile and confidence [5].
- **Enhanced Comfort:** Unlike removable dentures, implants are fixed in place, eliminating the discomfort associated with removable appliances.
- **Durability:** With proper care, dental implants can last a lifetime, making them a cost-effective long-term solution.
- **Bone Preservation:** Implants stimulate the jawbone, preventing bone loss and maintaining the facial structure.
- **Improved Oral Health:** Implants do not require the reduction of adjacent teeth, preserving natural tooth structure.

Challenges and Considerations, while dental implants offer numerous benefits, there are also challenges and considerations to keep in mind:

- **Surgical Risks:** The implant placement procedure involves surgery, which carries risks such as infection, nerve damage, and implant failure.
- **Cost:** Dental implants are generally more expensive than other tooth replacement options, although their longevity can justify the initial investment.
- **Bone Health:** Successful implantation requires sufficient bone density and volume. In cases of significant bone loss, additional procedures like bone grafting may be necessary.
- **Time:** The process of getting dental implants can take several months, from initial consultation and surgery to final restoration [6].

Advancements in Dental Implant Technology

Advancements in dental implant technology are continually improving the outcomes for patients. Innovations include:

Enhanced Materials: Development of new materials and composites, particularly in the resins used for prosthetic crowns, to improve durability, aesthetics, and biocompatibility.

Improved Techniques: Advancements in surgical techniques and imaging technology, such as computer-guided implant surgery, increase the precision and success rates of implant placement [7].

Surface Modifications: Treatments to the surface of the implant fixtures to enhance osseointegration (the process by which the implant integrates with the bone), leading to better stability and longevity.

Dental implants have become a cornerstone of modern restorative dentistry, offering reliable and natural-looking solutions for tooth loss [8]. Understanding the components, benefits, challenges, and ongoing advancements in dental implant technology is essential for both dental professionals and patients seeking effective and lasting tooth replacement options.

1.2 Historical development in dental implant technology

The history of dental implant technology is a testament to human ingenuity and the relentless pursuit of improving oral health. From ancient attempts to modern technological advancements, the evolution of dental implants highlights significant milestones that have transformed restorative dentistry. The earliest evidence of dental implants dates back to ancient civilizations, such as the Mayans and Egyptians, who used materials like shells, ivory, and carved stones to replace missing teeth. Archaeological findings reveal that these primitive implants often integrated well with the bone, indicating a basic understanding of biocompatibility.

In the 19th century, dentists experimented with gold and various alloys for dental implants. These materials were chosen for their durability and resistance to corrosion. However, these early attempts often faced challenges with integration and stability. In the early 20th century, Dr. Greenfield introduced a hollow cylinder made of iridio-platinum, and Dr. Alvin Strock used orthopedic screws, which marked significant steps toward modern implantology. These implants provided better structural support and integration with the bone. The pivotal moment in dental implant technology came in the 1950s with the work of Swedish orthopedic surgeon Per-Ingvar Brånemark. While conducting research on bone healing, Brånemark discovered that titanium could effectively bond with bone tissue, a process he termed "osseointegration." This discovery laid the foundation for modern dental implants. In 1965, Brånemark placed the first titanium dental implant in a human volunteer, which successfully functioned for several decades. This breakthrough led to the commercialization of titanium implants, revolutionizing restorative dentistry. Researchers focused on enhancing the surface properties of titanium implants to promote better osseointegration. Techniques such as acid etching, plasma spraying, and grit blasting were introduced to create micro-rough surfaces that encouraged bone growth. In the late 20th century, zirconia implants emerged as an alternative to titanium. Zirconia, a ceramic material, offered excellent biocompatibility, strength, and aesthetic advantages due to its tooth-like color.

The integration of digital technology has revolutionized dental implantology. Computer-guided surgery allows for precise planning and placement of implants, reducing surgical errors and improving outcomes. Advanced imaging techniques, such as cone beam computed tomography (CBCT), provide detailed views of the patient's oral anatomy. The application of nanotechnology has led to the development of implants with nanoscale surface modifications. These enhancements promote rapid osseointegration and reduce healing times. Innovations in biomaterials have produced coatings that release bioactive molecules to accelerate bone healing and reduce the risk of infection. These coatings can include antibiotics, growth factors, and other bioactive substances. Advances in implant stability and surgical techniques have enabled immediate loading protocols, where the prosthetic crown is placed shortly after implant insertion, reducing treatment time and improving patient satisfaction. The historical development of dental implant technology reflects a journey of continuous innovation and improvement. From ancient attempts with rudimentary materials to the sophisticated implants of today, each advancement has contributed to the high success rates and widespread acceptance of dental implants as a preferred solution for tooth replacement [9].

1.3 Comparison with traditional tooth replacement methods

Dental implants offer significant advantages over traditional tooth replacement methods in terms of stability, longevity, bone preservation, and aesthetics. While traditional methods are less invasive and have lower upfront costs, their long-term drawbacks make dental implants a more sustainable and functional solution for tooth replacement. Traditional tooth replacement methods like dentures, bridges, and partial dentures offer various benefits, they generally fall short compared to dental implants in terms of stability, functionality, bone preservation, and aesthetics. However, they remain viable options due to their lower cost and less invasive nature.

- **Dentures:** Removable prosthetics replacing multiple or full arches of teeth.
- **Dental Bridges:** Fixed prosthetics bridging gaps between missing teeth, supported by adjacent teeth.
- **Partial Dentures:** Removable devices replacing one or several teeth, supported by natural teeth and gums.

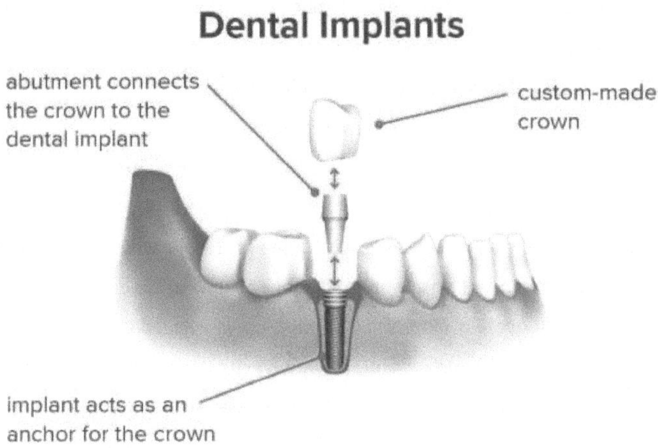

Fig. 1. View of Dental Implant

Dental implants are anchored into the jawbone, providing a stable and secure foundation for artificial teeth, mimicking the function of natural tooth roots. This stability allows for better chewing efficiency and clearer speech compared to dentures which can slip and cause discomfort. With proper care, dental implants can last a lifetime, making them a durable solution compared to dentures and bridges which may require replacements every few years. Made from biocompatible materials like titanium, implants resist wear and tear, ensuring long-term performance [10]. Dental implants stimulate the jawbone, preventing bone resorption and maintaining the natural shape of the face. This is a significant advantage over traditional dentures and bridges which do not support bone health. The osseointegration process where the implant fuses with the bone promotes bone growth and stability. Implants are designed to look, feel, and function like natural teeth, providing a seamless aesthetic. Unlike dentures, which can cause sore spots and discomfort, implants are custom-fitted to the mouth, ensuring comfort and reducing the risk of irritation. Implants require similar care to natural teeth, including regular brushing and flossing, without the need for special adhesives or cleaning solutions as with dentures. Unlike removable dentures, implants are fixed in place, eliminating the hassle of removal and daily reinsertion. Unlike bridges, which require adjacent teeth to be filed down, implants do not affect neighboring teeth, maintaining overall oral health. Properly placed implants can reduce the risk of gum disease and other oral health problems associated with gaps from missing teeth. Dental implants provide significant advantages over traditional tooth replacement methods in terms of stability, functionality, longevity, bone preservation, aesthetics, comfort, and convenience. They offer a durable, long-term solution that enhances overall oral health and quality of life.

1.4 Common materials used in dental implants

Dental implants primarily use materials like titanium and zirconia due to their excellent biocompatibility, strength, and integration with bone. Ceramic coatings and polymeric materials are also utilized to enhance specific properties and address aesthetic and allergenic concerns. These materials ensure durability, functionality, and aesthetic appeal, making dental implants a reliable solution for tooth replacement.

Titanium: Pure titanium and titanium alloys (e.g., Ti-6Al-4V) highly biocompatible, strong, lightweight, and resistant to corrosion. Promotes osseointegration (bone fusion), widely used and extensively studied with a high success rate.

Zirconia: Ceramic material, white color, high strength, and excellent biocompatibility. Aesthetic appeal (tooth-colored), hypoallergenic, and low plaque affinity. Suitable for patients with metal allergies and for those seeking metal-free implants.

Ceramic Coatings: Often applied to titanium implants to enhance biocompatibility and aesthetics. Improved integration with bone and soft tissues, aesthetic benefits, and reduced inflammation.

Polymeric Materials: Used in implant components like abutments and provisional restorations. Flexibility, ease of manufacturing, and reduced stress on surrounding bone.

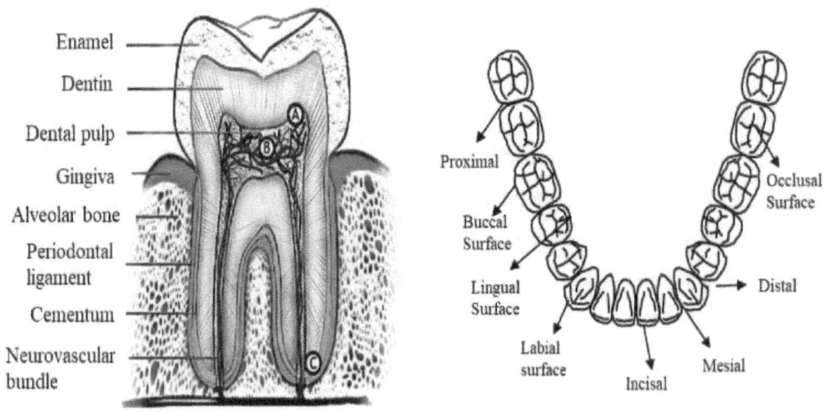

Fig. 2. Image of terminologies related to tooth geometry & tooth surfaces

1.5 Purpose of study

The primary purpose of this study is to enhance the material properties of resins used in dental implants to improve their overall performance. This includes, Improving Mechanical Properties which increases the strength, durability, and wear resistance of resin materials to ensure long-lasting and reliable dental implants. Ensuring the resin materials are compatible with biological tissues, reducing the risk of inflammation and rejection. Developing resins that better match the natural appearance of teeth, providing more visually appealing implants. Enhancing the functionality of dental implants to improve patient comfort and chewing efficiency. Utilizing advanced manufacturing techniques such as 3D printing to create more precise and custom-fitted dental implants. By achieving these objectives, the study aims to develop dental implants that not only meet but exceed current standards in terms of performance, safety, and patient satisfaction.

1.6 Expected outcome of the proposed work

Development of resin materials with improved strength, durability, and wear resistance, leading to longer-lasting dental implants. Creation of resin implants that are more compatible with biological tissues, reducing inflammation and rejection rates. Production of resin materials that closely mimic the natural appearance of teeth, resulting in more visually appealing implants. Dental implants that provide better comfort, stability, and chewing efficiency for patients. Utilization of advanced manufacturing methods, such as 3D printing, to create custom-fitted and precise dental implants. Enhanced patient satisfaction due to the improved performance, aesthetics, and longevity of the dental implants. Development of novel resin-based materials specifically designed for dental implants. These resins would ideally possess superior properties compared to existing options, including significantly exceeding the current brittleness of resins to withstand biting forces and prevent fracture. Promoting stronger osseointegration, the fusion of the implant with the jawbone, for long-term stability. Minimizing implant degradation over time, leading to a longer lifespan. Integration of these new resins into the design and functionality of dental implants. Creating micro or nano-textures on the implant surface to further enhance osseointegration. Developing intricate networks of pores within the resin to encourage bone ingrowth and improve implant stability. Designing implant shapes optimized for specific applications and patient needs. The expected outcome of this thesis is to create a new generation of dental implants made from advanced resins that offer significant improvements over existing options. This could lead to a future with more reliable, long-lasting, and patient-friendly dental implant procedures.

REFERENCES

[1] Farooq et al., “Synergistic effect of bioactive inorganic fillers in enhancing properties of dentin adhesives—a review,” Polymers (Basel)., vol. 13, no. 13, pp. 1–15, 2021, doi: 10.3390/polym13132169.

[2] M. M. Gad, S. M. Fouda, F. A. Al-Harbi, R. Näpänkangas, and A. Raustia, “PMMA denture base material enhancement: A review of fiber, filler, and nanofiller addition,” Int. J. Nanomedicine, vol. 12, pp. 3801–3812, 2017, doi: 10.2147/IJN.S130722.

[3] D. L. Yang, Q. Sun, H. Niu, R. L. Wang, D. Wang, and J. X. Wang, “The properties of dental resin composites reinforced with silica colloidal nanoparticle clusters: Effects of heat treatment and filler composition,” Compos. Part B Eng., vol. 186, no. September 2019, p. 107791, 2020, doi: 10.1016/j.compositesb.2020.107791.

[4] F. Watari, A. Yokoyama, F. Saso, M. Uo, and T. Kawasaki, “Fabrication and properties of functionally graded dental implant,” Compos. Part B Eng., vol. 28, no. 1–2, pp. 5–11, 1997, doi: 10.1016/s1359-8368(96)00021-2.

[5] S. G. Chen, J. Yang, Y. G. Jia, B. Lu, and L. Ren, “Tio2 and PEEK reinforced 3d printing pmma composite resin for dental denture base applications,” Nanomaterials, vol. 9, no. 7, 2019, doi: 10.3390/nano9071049.

[6] M. Silva, R. Felismina, A. Mateus, P. Parreira, and C. Malça, “Application of a Hybrid Additive Manufacturing Methodology to Produce a Metal/Polymer Customized Dental Implant,” Procedia Manuf., vol. 12, no. December 2016, pp. 150–155, 2017, doi: 10.1016/j.promfg.2017.08.019.

[7] J. B. Brunski, “Biomaterials and biomechanics in dental implant design.,” Int. J. Oral Maxillofac. Implants, vol. 3, no. 2, pp. 85–97, 1988, [Online]. Available: http://www.ncbi.nlm.nih.gov/pubmed/3075195.

[8] C. Blackburn, H. Rask, and A. Awada, “Mechanical properties of resin-ceramic CAD-CAM materials after accelerated aging,” J. Prosthet. Dent., vol. 119, no. 6, pp. 954–958, 2018, doi: 10.1016/j.prosdent.2017.08.016.

[9] D. Khorsandi et al., “3D and 4D printing in dentistry and maxillofacial surgery: Printing techniques, materials, and applications,” Acta Biomater., vol. 122, pp. 26–49, 2021, doi: 10.1016/j.actbio.2020.12.044.

[10] A. Dhakne, S. Jaju, and S. Shukla, “Review on analysis of enhancing wear properties through thermo-mechanical treatment and grain size,” Mater. Today Proc., vol. 60, pp. 2270–2272, Jan. 2022, doi: 10.1016/J.MATPR.2022.04.719.

2. LITERATURE REVIEW

Dental implants are a critical innovation in modern dentistry, providing reliable solutions for tooth replacement. The success of dental implants heavily relies on the material properties of their components, particularly resins. Resins are favored for their aesthetic qualities and versatility, but enhancing their material properties is essential for improving the longevity and functionality of dental implants. This review explores the progression of research in this area, identifying key advancements and ongoing challenges. Dental implants are a critical innovation in modern dentistry, providing reliable solutions for tooth replacement. The success of dental implants heavily relies on the material properties of their components, particularly resins. Resins are favored for their aesthetic qualities and versatility, but enhancing their material properties is essential for improving the longevity and functionality of dental implants [1]. This review explores the progression of research in this area, identifying key advancements and ongoing challenges. Dental implants are a critical innovation in modern dentistry, providing reliable solutions for tooth replacement. The success of dental implants heavily relies on the material properties of their components, particularly resins. Resins are favored for their aesthetic qualities and versatility, but enhancing their material properties is essential for improving the longevity and functionality of dental implants. This review explores the progression of research in this area, identifying key advancements and ongoing challenges.

The trend towards personalized dental implants using 3D printing and advanced resin formulations has gained momentum. Studies like Kim et al. (2021) highlight the potential of 3D-printed resins to offer customized solutions tailored to individual patient needs. The design of dental implants has evolved to improve the integration and performance of resin materials. Innovations such as tapered implants and surface modifications have enhanced the clinical outcomes of resin-based implants. Advances in resin technology, including the development of nanocomposites and bioactive resins, have significantly improved the mechanical and biological properties of these materials. Comparative studies have shown that modern resins can match or even surpass the performance of traditional materials like metals and ceramics in certain dental applications [2]. Changes in implant design have necessitated corresponding improvements in resin materials to ensure compatibility and enhanced performance.

2.1 Year wise/Author wise/Sub topic wise review

Fibre-reinforced composites (FRCs) that are continuous and unidirectional have been employed in many therapeutic appliances, including periodontal splints, orthodontic retainers, two to four-unit bridges for tooth restoration, and bone defects repair in orthopaedics (2020, Material Science and engineering Journal, Ting Wanga et al.). In many medical fields, including orthopaedics and dentistry, implants are now a common way to replace or support damaged body components. Nevertheless, even with advancements in surgery and implant technology, implant failure is still a major worry [3]. This comprehensive analysis attempts to provide an overview of the materials used in dental implants, the synthesis process, the reasons behind implant failure, and the associated clinical implications (2023, Results in Chemistry, Recent development of dental implant materials, synthesis process, and failure, Nayem Hossain et al.). Since bisphenol A (BPA) is a raw ingredient used in the synthesis of dental composites and has been linked to a number of health issues due to its estrogenic potential, the usage of BisGMA as the base monomer in these materials has come under scrutiny. In order to construct experimental photopolymerizable composite resins, this work evaluates the trimethacrylate tris (4-hydroxyphenyl) methane triglycidyl methacrylate (TTM) monomer and explains its manufacture (2020, Preparation and evaluation of a BisGMA-free dental composite resin based on a novel trimethacrylate monomer, Alma Antonia Pérez-Mondragóna et al.).

The degree of conversion and mechanical properties of Bis-GMA- and UDMA-based resin composites containing the fillers were examined in this study as a result of the surface conditioning of nano-zirconia fillers with 10-methacryloyloxydecyl dihydrogen phosphate (MDP) and dipentaerythritol penta-acrylate phosphate (PENTA) (2020, Journal of Dentistry, Jiaxue Yanga et al.). Dental resin composites, which are now an excellent substitute for metal and amalgam restorations, have transformed dental care by enabling minimally invasive dentistry to preserve healthy tooth structure and produce aesthetic outcomes that look natural [4]. However, in order to improve their durability and long-term clinical efficacy, dental composites are currently undergoing additional development. By thoroughly comprehending the diverse material attributes and design approaches of dental composite systems, dental restoration research can progress towards the development of innovative materials that yield properties, functions, and internal structures strikingly similar to those of hard dental tissues (2022, Composites Part B Engineering, Kiho Cho et al.). The mechanical characteristics, degree of conversion, flowability, and low AR (50/70/100) micro-sized fibres with a diameter of 5 μm were assessed for a variety of fiber-reinforced composites. The test programme in the publication closely followed ISO guidelines for statistical reliability and came to the conclusion that adding micro-sized fibres to dental composites for reinforcement can improve their flexural and compressive properties without appreciably changing their flowability or degree of conversion. Furthermore, the insertion of fibres decreased deformation prior to failure (2021, Materials Science & Engineering C, Sonam Behla et al.).

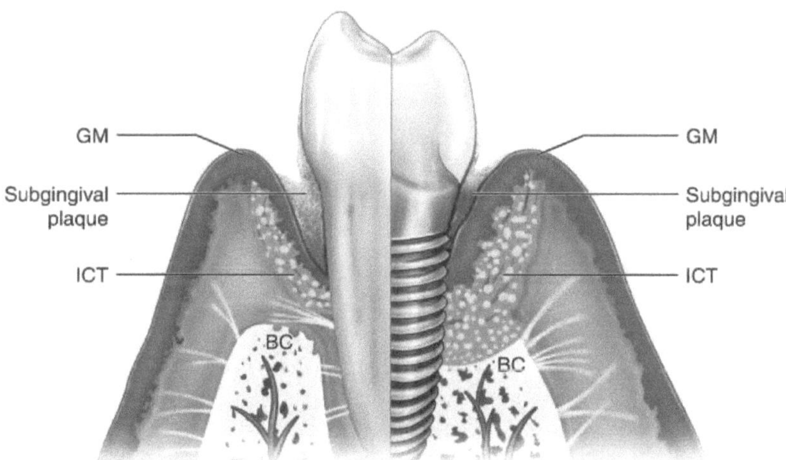

Fig. 3. Image of periodontitis (LHS) and peri-implantitis (RHS)

2.2 Gap Identification

Most studies provide short- to medium-term data on the performance of resin-based dental implants. There is a lack of long-term clinical studies (spanning over 10 years) that evaluate the durability, biological integration, and patient satisfaction with resin-based implants. Long-term studies are essential to determine the true efficacy and longevity of these materials in diverse patient populations [5]. Various studies use different methodologies for testing the mechanical properties and biocompatibility of resin materials. There is no standardized protocol for evaluating the properties of dental resin materials, making it difficult to compare results across different studies. Establishing standardized testing methods would facilitate more consistent and reliable comparisons, aiding in the development of superior resin materials. While initial biocompatibility studies are common, they often focus on limited aspects such as cytotoxicity or initial tissue response. There is a need for comprehensive biocompatibility studies that assess the long-term interaction of resin materials with oral tissues, including potential inflammatory responses and the impact on oral microbiota. Understanding the long-term biocompatibility of these materials is crucial for ensuring their safety and effectiveness in clinical applications. Some studies have started exploring the antimicrobial properties of resins, but there is limited research on how these materials affect the overall oral microbiome.

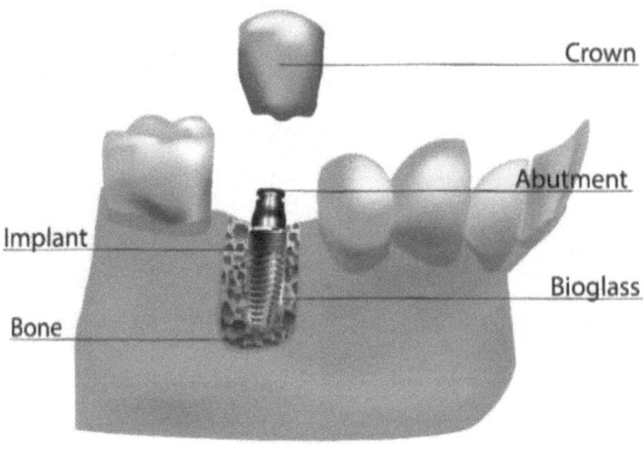

Fig. 4. Image of dental implant placed in oral cavity

There is a lack of in-depth studies examining the impact of resin materials on the oral microbiome, including potential shifts in microbial populations and the implications for oral health. Investigating these effects could lead to the development of resin materials that promote a healthy oral environment, reducing the risk of infections and other complications. Recent research has begun exploring smart materials and advanced resin formulations that can respond to environmental changes or release therapeutic agents. There is limited experimental and clinical data on the performance of these advanced materials, including their practical applications and benefits in real-world scenarios. More research is needed to validate the effectiveness of these materials and understand their potential in enhancing the functionality and longevity of dental implants. Many studies focus on the technical and material properties of resin-based implants, often neglecting patient-reported outcomes and experiences [6]. There is a scarcity of research that prioritizes patient feedback and evaluates the subjective experiences of individuals using these implants. Incorporating patient-centric research could provide valuable insights into the practical benefits and challenges of using resin-based implants, guiding improvements that enhance patient satisfaction.

The literature primarily focuses on the scientific and clinical aspects of resin-based dental implants, with limited discussion on the economic implications and accessibility for diverse populations. There is a need for studies that assess the cost-effectiveness and accessibility of advanced resin materials, considering different healthcare systems and socioeconomic factors. Addressing these issues is essential for ensuring that advancements in dental implant technology are available and affordable to a broader range of patients. Identifying and addressing these gaps is critical for the continued development of effective and durable dental implants [7]. Future research should focus on long-term performance, standardized testing, comprehensive biocompatibility, the impact on the oral microbiome, advanced material formulations, patient-centric outcomes, and economic accessibility to drive the field forward and improve patient care.

2.3 Possible Contribution

Conduct longitudinal studies that track the performance of resin-based dental implants over a period of 10-20 years. Conduct longitudinal studies that track the performance of resin-based dental implants over a period of 10-20 years. Use standardized protocols for patient selection, implant placement, and follow-up assessments to ensure consistent and reliable results. Develop and promote standardized testing protocols for evaluating the mechanical properties and biocompatibility of dental resin materials. Enable more accurate comparisons between different studies, facilitating the identification of the best-performing materials. Collaborate with international dental organizations and research institutions to create consensus guidelines and testing standards. Perform in-depth biocompatibility studies that investigate the long-term interaction of resin materials with oral tissues and the immune response. Ensure the safety and effectiveness of resin-based dental implants by identifying any potential adverse reactions. Use advanced imaging techniques, histological analysis, and molecular biology methods to assess tissue response and inflammatory markers. Research the effects of resin materials on the oral microbiome, including potential shifts in microbial populations and overall oral health.

Develop resin materials that promote a healthy oral environment, reducing the risk of infections and other complications [8]. Conduct microbiome sequencing studies and in vitro experiments to assess microbial changes and interactions with resin materials. Innovate new resin formulations that incorporate smart materials capable of responding to environmental changes or releasing therapeutic agents. Enhance the functionality and longevity of dental implants, offering benefits such as self-healing properties or antimicrobial effects. Combine materials science with bioengineering techniques to create and test these advanced formulations in both laboratory and clinical settings. Prioritize research that incorporates patient feedback and evaluates subjective experiences with resin-based dental implants. Improve patient satisfaction and outcomes by addressing practical concerns and preferences. Use surveys, interviews, and patient-reported outcome measures to gather comprehensive feedback and incorporate it into the design and development process [9]. Assess the cost-effectiveness and accessibility of advanced resin materials, considering different healthcare systems and socioeconomic factors. Ensure that advancements in dental implant technology are affordable and accessible to a wider range of patients. Conduct economic analyses and accessibility studies to evaluate the financial impact and availability of new materials in various regions. Foster interdisciplinary collaboration between materials scientists, bioengineers, clinicians, and dental practitioners. Accelerate innovation and translation of research findings into practical applications [10]. Organize workshops, conferences, and collaborative research projects to share knowledge and develop integrated solutions. By addressing these contributions, researchers and clinicians can significantly advance the field of dental implants, improving material properties of resins, patient outcomes, and overall oral health. These efforts will not only enhance the performance and durability of dental implants but also ensure their accessibility and acceptability among diverse patient populations.

2.4 Problem Definition

Dental implants are widely used in modern dentistry to replace missing teeth. They typically consist of a titanium post surgically inserted into the jawbone, acting as an artificial tooth root, and a crown that mimics the visible part of a natural tooth. While titanium and its alloys are the most common materials for the implant post due to their excellent biocompatibility and mechanical properties, there are significant challenges associated with the materials used for the crown and abutment. Traditional materials used in dental implants, such as ceramics and metal alloys, often face challenges in terms of brittleness and fatigue resistance. These limitations can lead to fractures or wear over time, reducing the longevity of the implant [11]. Ceramic materials, while aesthetically pleasing, can be prone to chipping and lack the same level of translucency and color matching as natural teeth. This can lead to suboptimal cosmetic outcomes. While titanium is biocompatible, the materials used in the abutment and crown must also support good osseointegration and tissue response. Poor material choices can lead to inflammation, infection, or failure of the implant. The manufacturing processes for traditional implant materials can be complex and costly, impacting the affordability and accessibility of dental implants.

REFERENCES

[1] T. Wang, J. P. Matinlinna, J. He, K. E. Ahmed, and M. F. Burrow, "Biomechanical and biological evaluations of novel BPA-free fibre-reinforced composites for biomedical applications," Mater. Sci. Eng. C, vol. 117, no. July, p. 111309, 2020, doi: 10.1016/j.msec.2020.111309.

[2] N. Hossain et al., "Recent development of dental implant materials, synthesis process, and failure – A review," Results Chem., vol. 6, no. July, p. 101136, 2023, doi: 10.1016/j.rechem.2023.101136.

[3] A. A. Pérez-Mondragón, C. E. Cuevas-Suárez, J. A. González-López, N. Trejo-Carbajal, M. Meléndez-Rodríguez, and A. M. Herrera-González, "Preparation and evaluation of a BisGMA-free dental composite resin based on a novel trimethacrylate monomer," Dent. Mater., vol. 36, no. 4, pp. 542–550, 2020, doi: 10.1016/j.dental.2020.02.005.

[4] J. A. González-López, A. A. Pérez-Mondragón, C. E. Cuevas-Suárez, N. Trejo-Carbajal, and A. M. Herrera-González, "Evaluation of dental composites resins formulated with non-toxic monomers derived from catechol," J. Mech. Behav. Biomed. Mater., vol. 104, no. November 2019, 2020, doi: 10.1016/j.jmbbm.2019.103613.

[5] J. A. González-López, A. A. Pérez-Mondragón, C. E. Cuevas-Suárez, S. C. Esparza González, and A. M. Herrera-González, "Dental composite resins with low polymerization stress based on a new allyl carbonate monomer," J. Mech. Behav. Biomed. Mater., vol. 110, no. April, pp. 1–7, 2020, doi: 10.1016/j.jmbbm.2020.103955.

[6] H. Chai, X. Wang, and J. Sun, "Miniature specimens for fracture toughness evaluation of dental resin composites," Dent. Mater., vol. 35, no. 2, pp. 283–291, 2019, doi: 10.1016/j.dental.2018.11.023.

[7] S. Behl, Raju, G. Rajan, A. Ellakwa, P. Farrar, and B. G. Prusty, "Physical and mechanical characterisation of flowable dental composites reinforced with short aspect ratio micro-sized S-Glass fibres," Mater. Sci. Eng. C, vol. 111, no. January, p. 110771, 2020, doi: 10.1016/j.msec.2020.110771.

[8] Y. Catel, J. Angermann, P. Fässler, U. Fischer, T. Schnur, and N. Moszner, "High refractive index monofunctional monomers as promising diluents for dental composites," Dent. Mater., vol. 37, no. 2, pp. 351–358, 2021, doi: 10.1016/j.dental.2020.11.029.

[9] K. Cho, G. Rajan, P. Farrar, L. Prentice, and B. G. Prusty, "Dental resin composites: A review on materials to product realizations," Compos. Part B Eng., vol. 230, no. January 2022, 2022, doi: 10.1016/j.compositesb.2021.109495.

[10] N. Jambhulkar, S. Jaju, and A. Raut, "Surface modification techniques for different materials used in dental implants review," Mater. Today Proc., vol. 60, pp. 2266–2269, Jan. 2022, doi: 10.1016/J.MATPR.2022.04.683.

[11] R. V. Dahibhate and S. B. Jaju, "Effect of irrigation mode, volume, medium and drill on heat generation during implant surgery: A review," Mater. Today Proc., vol. 50, pp. 1799–1803, Jan. 2022, doi: 10.1016/J.MATPR.2021.09.206.

3. FOCUS ON RESINS IN DENTAL IMPLANTS

The goal of this research and development project is to enhance the material properties of dental implants, particularly focusing on advanced resin materials that can address the limitations mentioned above. Dental implants are crucial for restoring the function and aesthetics of missing teeth. Traditional materials used in dental implants include metals (such as titanium), ceramics, and polymers. Resins, particularly advanced dental resins, have emerged as promising materials due to their potential for customization, aesthetic appeal, and improved mechanical properties. However, several challenges must be addressed to fully utilize resins in dental implants. Conventional resins often lack the necessary strength and durability, leading to potential failure under the high-stress conditions of the oral environment. Resins are prone to wear and abrasion, which can compromise the longevity of the implant and the adjacent natural teeth. Achieving a natural look with resins can be challenging due to issues with translucency, color stability, and the ability to mimic the optical properties of natural teeth. Ensuring that resin materials are biocompatible to prevent adverse tissue reactions and promote osseointegration is critical. Poor biocompatibility can lead to inflammation, infection, and implant failure. Resins can absorb water and other oral fluids, which can lead to dimensional changes, degradation, and decreased mechanical properties over time [1]. Developing resins that are easy to manufacture, shape, and process while maintaining high performance and quality is a significant challenge.

Develop resin materials with improved tensile strength, fracture resistance, and wear resistance. Formulate resins that provide superior color stability, translucency, and the ability to mimic natural tooth enamel. Design resin materials that are biocompatible, reducing the risk of adverse tissue reactions and promoting healthy integration with surrounding tissues. Create resin formulations that are resistant to water absorption and solubility to maintain their mechanical integrity over time. Develop efficient and cost-effective manufacturing processes for resin-based dental implants. Identify and analyze various resin formulations and additives to evaluate their mechanical, aesthetic, and biocompatible properties [2]. Experiment with composites and nano-fillers to enhance the properties of the resins. Conduct in vitro and in vivo testing to assess the performance, durability, biocompatibility, and aesthetic quality of the developed resin materials.

3.1 Overview of resin materials used in dental applications

Resin materials have become a cornerstone in modern dental applications due to their versatility, aesthetic properties, and ease of use. They are widely used in various dental procedures, including restorations, prosthetics, and orthodontics [3]. This overview highlights the different types of resin materials, their properties, applications, and recent advancements.

Composite Resins, Typically consist of a matrix of organic polymer (usually bis-GMA or UDMA) and inorganic filler particles (such as silica or zirconia). High aesthetic value, good mechanical properties, and improved wear resistance. They can be tailored for different applications by varying the filler content and size. Acrylic Resins, Made from polymethyl methacrylate (PMMA) and its derivatives. Good aesthetics, ease of manipulation, and cost-effectiveness. However, they may have lower mechanical strength compared to composite resins. Denture bases, temporary crowns, and orthodontic appliances. Polyetheretherketone (PEEK) Resins, High-performance thermoplastic polymer [4]. Excellent mechanical strength, biocompatibility, and resistance to wear and chemical attack. They are radiolucent and have a similar modulus of elasticity to bone. Frameworks for removable and fixed prosthetics, implant abutments, and partial dentures. Glass Ionomer Cements (GICs), Consist of fluoroaluminosilicate glass and polyacrylic acid. Chemical bond to tooth structure, release fluoride, and have good biocompatibility. However, they have lower aesthetic properties and mechanical strength compared to composite resins. Luting cements, cavity liners, base materials, and restorative materials for non-load-bearing areas. Hybrid Resins, Combine properties of composite resins and glass ionomers. Improved mechanical strength and aesthetics compared to GICs, with the added benefit of fluoride release. Restorative materials in areas requiring both strength and fluoride release. Incorporation of nano-sized fillers has improved the mechanical properties and aesthetics of composite resins, providing better wear resistance and polishability. Development of resins that can release ions to promote remineralization and inhibit bacterial growth, enhancing the longevity of restorations. Integration of computer-aided design and manufacturing (CAD/CAM) has enabled the precise fabrication of resin-based restorations, improving fit and function. Advances in 3D printing technologies have allowed for the custom fabrication of resin-based dental appliances and restorations with high precision and speed [5]. Resin materials play a crucial role in modern dentistry due to their versatility, aesthetic appeal, and functional properties. Continuous advancements in resin formulations and technologies are enhancing their performance and expanding their applications, leading to better patient outcomes and more efficient dental practices.

3.2 Role of materials in the success and longevity of dental implants

The success and longevity of dental implants are significantly influenced by the materials used in their construction. These materials must meet stringent requirements to ensure durability, biocompatibility, mechanical stability, and aesthetic appeal. Biocompatible materials minimize the risk of adverse reactions, such as inflammation, infection, and allergic responses. They promote proper osseointegration, which is the direct structural and functional connection between living bone and the surface of the implant. Zirconia: A ceramic material known for its biocompatibility and aesthetic qualities, often used for implant abutments and crowns. Dental implants must withstand significant mechanical forces from chewing and biting without fracturing or deforming. They should have appropriate tensile strength, fatigue resistance, and fracture toughness. Polyetheretherketone (PEEK): A high-performance polymer used for its strength and elasticity, similar to natural bone, often in implant frameworks [6]. Aesthetics are crucial, especially for visible teeth. The material should mimic the appearance of natural teeth in terms of color, translucency, and texture. Zirconia: Highly aesthetic with a natural tooth-like appearance, used for crowns and abutments. Materials should resist wear and corrosion from the acidic and enzymatic environment of the mouth to maintain structural integrity over time. Successful osseointegration ensures that the implant becomes securely anchored to the jawbone, providing stability and preventing implant failure.

Bioactive Glass and Coatings can be applied to implants to promote bone bonding and enhance osseointegration. The material must be amenable to precise manufacturing techniques to ensure proper fit, function, and aesthetic outcomes. Resins (for 3D printing) allow for the custom fabrication of dental components with high accuracy and reduced manufacturing time.

Nanostructured Titanium improves osseointegration and reduces the risk of bacterial colonization. The success and longevity of dental implants are heavily dependent on the selection and application of appropriate materials [7]. These materials must provide a combination of biocompatibility, mechanical strength, aesthetic appeal, and resistance to wear and corrosion. Advances in materials science, such as the development of nanomaterials and bioactive coatings, continue to enhance the performance of dental implants, leading to better patient outcomes and longer-lasting solutions.

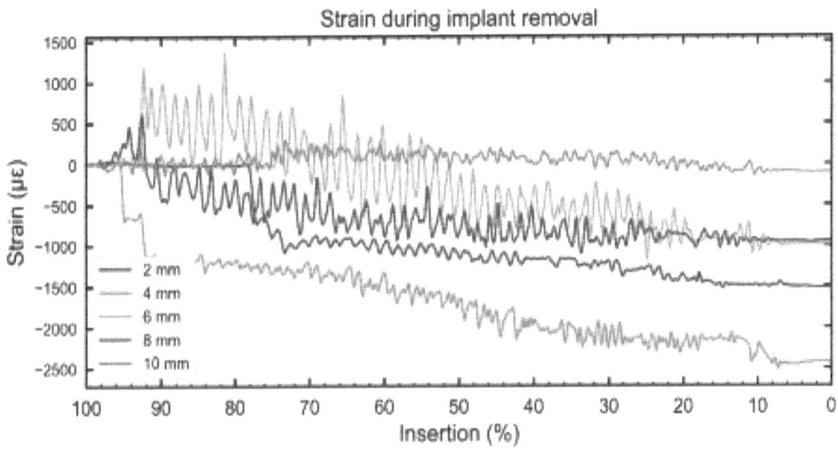

Fig. 5. Image of strain during implant removal

3.3 Significance of Material Properties in Dental Implants

The material properties of dental implants play a crucial role in their success, functionality, and longevity. Each property contributes to how well the implant performs in the oral environment, addressing both mechanical and biological challenges. Biocompatibility, ensures that the implant does not provoke an immune response or cause inflammation and infection. Mechanical Strength and Durability, Implants must withstand substantial mechanical forces from chewing and biting without deforming or fracturing, high mechanical strength ensures long-term functionality and reduces the likelihood of implant failure. Fatigue Resistance implants are subjected to repeated cyclic loading during chewing, which can lead to material fatigue and failure over time, High fatigue resistance ensures that the implant can endure these repeated stresses without failing. Aesthetic Properties, are essential for implants in visible areas of the mouth to ensure a natural appearance. Materials must match the color, translucency, and texture of natural teeth. Wear and Corrosion Resistance, implants are exposed to the harsh oral environment, including acidic conditions and enzymatic activity, which can cause wear and corrosion. Resistance to wear and corrosion ensures the longevity and functionality of the implant [8]. Osseointegration is the ability of the implant material to bond with the bone is critical for stability and preventing implant failure. Good osseointegration leads to a stable and secure implant, reducing the risk of loosening and failure. Thermal and Electrical Conductivity materials should have low thermal conductivity to prevent damage to surrounding tissues from temperature changes. Electrical conductivity can influence the biological response and osseointegration. Ease of Manufacturing and Processing materials must be easy to fabricate and customize to fit the patient's specific anatomy. Advances in manufacturing technologies like CAD/CAM and 3D printing rely on materials that can be precisely shaped and processed. Bioactive materials can interact positively with biological tissues, promoting healing and integration [9]. They can release beneficial ions that aid in tissue regeneration and reduce the risk of infection.

The material properties of dental implants are crucial for their success and longevity. Each property, from biocompatibility and mechanical strength to aesthetic qualities and wear resistance, contributes to the overall performance of the implant. Advances in material science continue to improve these properties, leading to better patient outcomes and more reliable dental implant solutions. Understanding and optimizing these properties are essential for developing effective and long-lasting dental implants.

3.4 Challenges associated with current materials

Current materials used in dental implants face several challenges that impact their performance and longevity. These challenges span across various aspects, from mechanical properties to biocompatibility and aesthetic integration. Mechanical Properties, Imagine a bridge enduring relentless traffic; similarly, dental implants must withstand the constant forces of chewing and biting. However, traditional materials like ceramics and metals may exhibit brittleness or fatigue over time, leading to fractures or wear [10]. This compromise in structural integrity poses a significant challenge to the longevity of the implant. Aesthetic Considerations, Just as an artist strives for perfection in their masterpiece, achieving natural aesthetics in dental implants is paramount. However, materials like ceramics, while visually appealing, may lack the translucency and color matching capabilities of natural teeth. This discrepancy can detract from the overall cosmetic outcome, posing a challenge in achieving seamless integration with the patient's smile. Biocompatibility and Osseointegration, Picture a puzzle missing a crucial piece; similarly, poor biocompatibility or inadequate osseointegration of implant materials can lead to inflammatory reactions or even outright rejection by the body. Ensuring that materials seamlessly integrate with surrounding tissues is essential for the long-term success of dental implants. Manufacturing and Cost, Imagine a complex jigsaw puzzle with numerous intricate pieces; similarly, manufacturing processes for traditional implant materials can be intricate and costly. This complexity can pose challenges in scalability and affordability, limiting accessibility to a broader population. Wear and Corrosion Resistance, Just as erosion wears away at rocks over time, the oral environment can subject dental implants to wear and corrosion. Traditional materials may struggle to withstand these harsh conditions, compromising their durability and longevity. Stability and Fit, Like a key fitting snugly into a lock, dental implants must be precisely fabricated and fitted to ensure stability and functionality. Challenges in achieving this precise fit can lead to issues such as improper seating or loosening over time. By developing materials that offer superior mechanical properties, enhanced biocompatibility, and improved aesthetics, researchers can pave the way for more durable and cosmetically pleasing dental implants. Additionally, streamlining manufacturing processes to reduce costs and increase accessibility will further contribute to overcoming these challenges and improving patient outcomes in implant dentistry [11].

3.5 Advantages of resins (aesthetics, ease of manipulation)

Resins offer several advantages in dental applications, particularly in terms of aesthetics and ease of manipulation. Natural Appearance, Resins can be color-matched to natural teeth, allowing for seamless integration with the patient's smile. This ability to mimic the appearance of natural teeth makes resin-based restorations virtually indistinguishable from the surrounding dentition, enhancing overall aesthetic outcomes. Translucency, Many resin materials exhibit translucency similar to natural tooth enamel, providing a lifelike appearance. This translucency allows light to pass through the restoration, creating a vibrant, natural-looking smile. Customization, Resins offer a high degree of customization, allowing dental professionals to create restorations that precisely match the shape, size, and color of the patient's existing teeth. This customization ensures that the restoration blends harmoniously with the surrounding dentition, enhancing aesthetic harmony. Versatility, Resins are highly versatile materials that can be used in various dental procedures, including fillings, veneers, crowns, and bridges. Their versatility allows dental professionals to address a wide range of dental issues using a single material, simplifying treatment planning and execution. Ease of Shaping and Sculpting, Resins can be easily shaped and sculpted during the restoration process, allowing dental professionals to achieve precise contours and anatomical features. This ease of manipulation enables the creation of restorations that not only look natural but also function effectively within the oral cavity. Minimal Tooth Preparation, In many cases, resin-based restorations require minimal tooth preparation compared to traditional materials like metal or ceramic. This conservative approach helps preserve more of the patient's natural tooth structure, leading to better long-term dental health outcomes.

3.6 Limitations of conventional resins

While conventional resins have numerous advantages, they also possess limitations that can impact their performance and suitability for certain dental applications. Limited Strength, Conventional resins may have lower mechanical strength compared to traditional materials like metal or ceramic. This limitation can lead to issues such as fracture or wear under high stress conditions, compromising the longevity of restorations. Some conventional resin materials are prone to brittleness, particularly in thin or delicate areas of the restoration. This brittleness can increase the risk of fracture or chipping, especially in areas of the mouth subject to significant biting forces. Conventional resins may exhibit higher wear rates compared to other materials like ceramics or metal alloys. This susceptibility to wear can result in the deterioration of the restoration over time, necessitating replacement or repair. Certain resin materials may be prone to staining or discoloration over time, particularly in the presence of dietary factors such as coffee, tea, or tobacco.

This aesthetic deterioration can diminish the appearance of the restoration and impact patient satisfaction. Conventional resin materials undergo polymerization shrinkage during the curing process, leading to the formation of gaps between the restoration and the tooth structure. These gaps can compromise the integrity of the restoration and increase the risk of secondary caries or microleakage. Some conventional resin materials have poor radiopacity, making them difficult to detect on dental radiographs. This limitation can hinder the accurate diagnosis of underlying issues or complications associated with the restoration. Conventional resin materials require precise technique and handling to achieve optimal results. Factors such as moisture control, bonding protocols, and proper curing can significantly influence the performance and longevity of resin-based restorations. While resin materials offer excellent aesthetic properties, they may have limitations in terms of color matching or translucency compared to other materials like ceramics. This limitation can impact the overall aesthetic outcome of the restoration, particularly in highly visible areas of the mouth. Some patients may be allergic or sensitive to certain components of resin materials, leading to adverse reactions or tissue irritation. Ensuring the biocompatibility of resin materials is essential to minimize the risk of such complications.

REFERENCES

[1] B. M. Fronza, S. Lewis, P. K. Shah, M. D. Barros, M. Giannini, and J. W. Stansbury, "Modification of filler surface treatment of composite resins using alternative silanes and functional nanogels," Dent. Mater., vol. 35, no. 6, pp. 928–936, 2019, doi: 10.1016/j.dental.2019.03.007.

[2] W. Fu, L. Wang, and J. He, "Evaluation of mechanical properties and shrinkage stress of thiol-ene-methacrylate dental composites with synthesized fluorinated allyl ether," J. Mech. Behav. Biomed. Mater., vol. 95, no. January, pp. 53–59, 2019, doi: 10.1016/j.jmbbm.2019.03.027.

[3] A. P. Fugolin et al., "Alternative monomer for BisGMA-free resin composites formulations," Dent. Mater., vol. 36, no. 7, pp. 884–892, 2020, doi: 10.1016/j.dental.2020.04.009.

[4] W. Luo, Q. Huang, F. Liu, Z. Lin, and J. He, "Synthesis of antibacterial methacrylate monomer derived from thiazole and its application in dental resin," J. Mech. Behav. Biomed. Mater., vol. 49, pp. 61–68, 2015, doi: 10.1016/j.jmbbm.2015.04.026.

[5] V. Melinte, T. Buruiana, A. Chibac, M. Mares, H. Aldea, and E. C. Buruiana, "New acid BisGMA analogs for dental adhesive applications with antimicrobial activity," Dent. Mater., vol. 32, no. 12, pp. e314–e326, 2016, doi: 10.1016/j.dental.2016.09.026.

[6] S. V. Palagummi, T. Hong, Z. Wang, C. K. Moon, and M. Y. M. Chiang, "Resin viscosity determines the condition for a valid exposure reciprocity law in dental composites," Dent. Mater., vol. 36, no. 2, pp. 310–319, 2020, doi: 10.1016/j.dental.2019.12.003.

[7] M. Par, N. Spanovic, D. Mohn, T. Attin, T. T. Tauböck, and Z. Tarle, "Curing potential of experimental resin composites filled with bioactive glass: A comparison between Bis-EMA and UDMA based resin systems," Dent. Mater., vol. 36, no. 6, pp. 711–723, 2020, doi: 10.1016/j.dental.2020.03.015.

[8] X. Wang et al., "High performance dental resin composites with hydrolytically stable monomers," Dent. Mater., vol. 34, no. 2, pp. 228–237, 2018, doi: 10.1016/j.dental.2017.10.007.

[9] R. Srivastava, J. Liu, C. He, and Y. Sun, "BisGMA analogues as monomers and diluents for dental restorative composite materials," Mater. Sci. Eng. C, vol. 88, no. 2017, pp. 25–31, 2018, doi: 10.1016/j.msec.2018.01.011.

[10] R. V. Dahibhate, S. B. Jaju, and R. I. Sarode, "Development of mathematical model for prediction of bone drilling temperature," Mater. Today Proc., vol. 38, pp. 2732–2736,

Jan. 2021, doi: 10.1016/J.MATPR.2020.08.537.

[11] K. Ravekar, S. B. Jaju, and A. Zoeb Narsingpurwala, "Comparative analysis of material for material discharge plate of microsurfacing machine," Mater. Today Proc., vol. 46, pp. 249–255, Jan. 2021, doi: 10.1016/J.MATPR.2020.07.633.

4. RESEARCH METHODOLOGY

Research methodologies in dental implants for enhancing the material properties of resins encompass a range of approaches aimed at investigating, developing, and optimizing resin-based materials for dental applications. Analytical techniques such as scanning electron microscopy (SEM), X-ray diffraction (XRD) and Fourier-transform infrared spectroscopy (FTIR) to characterize material composition and structure. Mechanical testing methods including tensile testing, compression testing, and hardness testing to assess mechanical properties. Surface analysis techniques like atomic force microscopy (AFM) and surface profilometry to evaluate surface roughness and topography. Chemical analysis to determine material degradation, water sorption, and solubility [1]. Polymer chemistry techniques such as copolymerization, cross-linking, and blending to modify resin matrices. Incorporation of reinforcing agents (e.g., nanoparticles, fibers) or additives (e.g., antimicrobial agents, antioxidants) to improve material properties. Optimization of curing protocols, resin monomer ratios, and filler particle size distribution to enhance material performance.

Finite element analysis (FEA) to simulate mechanical loading conditions and predict stress distribution, deformation, and failure modes. Molecular modeling and simulations to investigate material interactions at the molecular level and predict material properties. Computational fluid dynamics (CFD) modeling to simulate fluid flow, erosion, and wear in oral environments. Optimization algorithms to design materials with specific mechanical, biological, and physicochemical properties. Long-term clinical follow-up studies to monitor implant survival, peri-implant health, and restoration performance in patient populations. Comparative clinical trials to evaluate the performance of resin-based implants versus traditional implant materials [2]. Collaboration with dental professionals and industry partners to facilitate technology transfer and commercialization of novel materials.

4.1 Properties of Resins in Dental Applications

Resins play a crucial role in various dental applications due to their unique properties, which make them suitable for a wide range of uses. Resins can be color-matched to natural teeth, allowing for seamless integration with the patient's dentition. Many resin materials exhibit translucency similar to natural tooth enamel, providing a lifelike appearance. Resins can be formulated with varying degrees of opacity to match different shades and translucencies of natural teeth. Resins can exhibit adequate flexural strength to withstand biting forces and chewing stresses. Some resin materials possess high compressive strength, making them suitable for load-bearing applications. Resins can have sufficient tensile strength to resist fracture and deformation under tension. Resins can form strong bonds with tooth structure, allowing for secure attachment of restorative materials [3]. Resins can create micro-mechanical interlocking with tooth surfaces, enhancing bond strength and retention. Resins are generally biocompatible with oral tissues, minimizing the risk of adverse reactions or tissue irritation.

Some resin materials promote favorable cellular response and tissue integration, facilitating healing and regeneration. Resins can be used in various dental procedures, including restorations, veneers, crowns, bridges, and orthodontic appliances. Resins offer a high degree of customization, allowing dental professionals to create tailored restorations to meet individual patient needs. Resins can be easily shaped and molded during the fabrication process, allowing for precise contouring and shaping [4]. Resin materials can be cured using different techniques (e.g., light-curing, chemical curing), providing control over the curing process. Resin-based restorations are often repairable, allowing for the easy correction of minor defects or damage without the need for replacement. Resins are often more cost-effective than alternative materials such as ceramics or metal alloys, making them a preferred choice for certain applications. Some resin materials possess sufficient radiopacity, allowing for easy detection on dental radiographs for diagnosis and monitoring.

4.2 Limitations of Conventional Resin Materials

While conventional resin materials offer numerous advantages in dental applications, they also possess certain limitations that can impact their performance and suitability for certain uses. Conventional resin materials may have lower mechanical strength compared to other materials like ceramics or metal alloys, making them less suitable for load-bearing applications. Some resin materials can be brittle, particularly in thin or delicate areas of restorations, increasing the risk of fracture or chipping. Conventional resin materials may exhibit higher wear rates compared to alternative materials, leading to deterioration of the restoration over time. Certain resin materials are prone to staining or discoloration over time, particularly in the presence of dietary factors like coffee, tea, or tobacco. Conventional resin materials undergo polymerization shrinkage during the curing process, leading to the formation of gaps between the restoration and the tooth structure [5]. This can compromise the integrity of the restoration and increase the risk of microleakage and secondary caries. Some conventional resin materials have poor radiopacity, making them difficult to detect on dental radiographs. This limitation can hinder the accurate diagnosis of underlying issues or complications associated with the restoration.

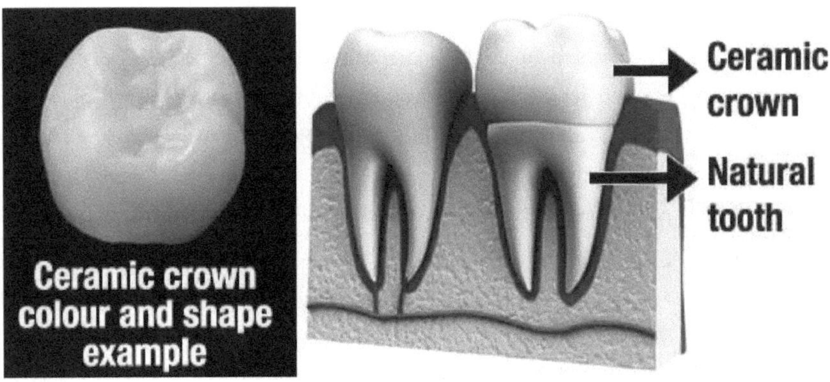

Fig. 6. Image of Ceramic Crown colour and its shape

Conventional resin materials require precise technique and handling to achieve optimal results. Factors such as moisture control, bonding protocols, and proper curing techniques can significantly influence the performance and longevity of resin-based restorations. While resin materials offer excellent aesthetic properties, they may have limitations in terms of color matching, translucency, or surface texture compared to other materials like ceramics. This limitation can impact the overall aesthetic outcome of the restoration, particularly in highly visible areas of the mouth. Some patients may be allergic or sensitive to certain components of resin materials, leading to adverse reactions or tissue irritation. Ensuring the biocompatibility of resin materials is essential to minimize the risk of such complications [6].

4.3 Nano-fillers and Surface Modification Techniques

Nano-fillers and surface modification techniques play significant roles in enhancing the properties and performance of resin materials in dental applications. Nano-fillers are nanoscale particles incorporated into resin matrices to improve various properties, such as mechanical strength, wear resistance, and aesthetic properties. Silica nanoparticles are widely used as reinforcing fillers in resin composites. They enhance mechanical properties such as flexural strength, modulus of elasticity, and hardness. Surface-modified silica nanoparticles can also improve dispersion within the resin matrix and reduce agglomeration. Titania nanoparticles are known for their photocatalytic properties and can enhance the antimicrobial properties of resin materials [7]. They also contribute to improved mechanical properties and wear resistance. Zirconia nanoparticles are used as reinforcing fillers in resin composites, providing improvements in mechanical properties such as flexural strength and fracture toughness. They also contribute to enhanced wear resistance and radiopacity. Calcium phosphate nanoparticles, such as hydroxyapatite or tricalcium phosphate, can improve the bioactivity and biocompatibility of resin materials. They promote remineralization of tooth structure and enhance bonding to natural tooth tissues. Silane coupling agents are molecules with both organic and inorganic functional groups that bond chemically to the surface of nanoparticles and the resin matrix.

They improve the interfacial adhesion between nano-fillers and the resin matrix, enhancing mechanical properties and wear resistance. Surface coatings, such as polymer coatings or surfactant layers, are applied to nano-fillers to improve their dispersion within the resin matrix and prevent agglomeration. These coatings can also enhance compatibility with the resin matrix and reduce filler-matrix interactions that may lead to degradation [8]. Functional groups, such as hydroxyl, amino, or carboxyl groups, can be introduced onto the surface of nano-fillers to modify their chemical properties and enhance interactions with the resin matrix. Functionalized nano-fillers improve bonding and dispersion within the composite, leading to improved mechanical and biological properties. Nano-fillers can be surface-modified or loaded with antibacterial agents, such as silver nanoparticles or quaternary ammonium compounds, to impart antimicrobial properties to resin materials. This surface modification inhibits the growth of bacteria and reduces the risk of secondary caries and periodontal disease.

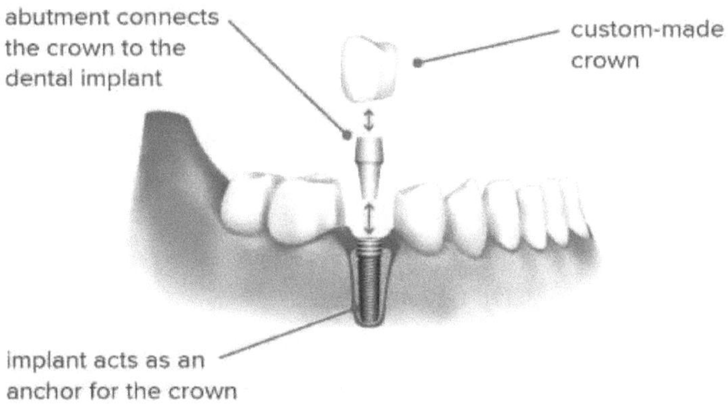

Fig. 7. Image of Crown to the dental implant

4.4 Development of Hybrid Resin Composites

The development of hybrid resin composites represents an innovative approach to optimizing the properties and performance of dental restorative materials. Hybrid resin composites combine the advantages of different filler types, resin matrices, and surface modification techniques to achieve a synergistic effect, resulting in enhanced mechanical, aesthetic, and clinical properties. Hybrid composites typically incorporate a combination of inorganic fillers such as silica, zirconia, titania, or glass ceramics. These fillers provide reinforcement and improve mechanical properties such as strength, wear resistance, and fracture toughness. Organic fillers, such as pre-polymerized resin particles or polymerizable monomers, may also be included to enhance flexibility, adaptability, and handling characteristics. The filler loading in hybrid composites is carefully optimized to achieve the desired balance between mechanical properties, aesthetic characteristics, and handling properties [9]. Higher filler loadings typically result in improved mechanical strength but may affect other properties such as translucency and polishability. Hybrid composites may incorporate fillers with varying particle sizes to optimize packing density, improve flowability, and minimize filler agglomeration. Controlled particle size distribution contributes to uniform distribution within the resin matrix and enhances overall composite performance. The inclusion of polymerization initiators and inhibitors in the resin matrix helps control the polymerization process, minimize shrinkage stress, and enhance polymerization kinetics.

These additives contribute to improved handling characteristics and clinical performance. Surface modification techniques, such as silane coupling agents, are employed to enhance the bonding between filler particles and the resin matrix. Silane coupling agents chemically bond to the surface of fillers and improve interfacial adhesion, resulting in enhanced mechanical properties and durability. Functional groups, such as hydroxyl or methacrylate groups, may be introduced onto the surface of fillers to enhance compatibility with the resin matrix and promote strong bonding. Functionalization improves filler dispersion, reduces agglomeration, and enhances composite performance. Hybrid composites are typically fabricated using conventional manufacturing methods such as bulk filling, incremental layering, or injection molding [10]. The choice of fabrication technique depends on factors such as restoration size, complexity, and clinical requirements. The curing protocol for hybrid composites is optimized to ensure complete polymerization and adequate mechanical properties. Factors such as light intensity, exposure time, and curing mode (e.g., light-curing, chemical curing) are carefully controlled to achieve optimal results. Hybrid composites undergo rigorous mechanical testing to evaluate properties such as flexural strength, compressive strength, hardness, wear resistance, and fracture toughness. These tests assess the material's ability to withstand occlusal forces and simulate long-term clinical performance. Hybrid composites are evaluated for biocompatibility using in vitro cytotoxicity assays, tissue response studies, and clinical trials to ensure patient safety and minimize adverse reactions. Clinical studies and trials are conducted to assess the performance of hybrid composites in real-world applications, including durability, esthetics, marginal integrity, and patient satisfaction [11]. Long-term clinical follow-up is essential to evaluate restoration longevity and clinical success rates.

4.4.1 In Vitro Testing Methods

Objective: Evaluate the performance of modified resin materials under controlled laboratory conditions.

Methods:

1. Mechanical testing to assess properties such as tensile strength, flexural strength, fracture toughness, and wear resistance.
2. Biocompatibility testing using cell culture assays (e.g., cytotoxicity assays, cell adhesion assays) to evaluate cellular response to materials.
3. Surface characterization techniques to examine changes in surface morphology, roughness, and chemical composition.
4. Aging studies to investigate material stability, degradation, and long-term performance.

4.4.2 In Vivo Testing Methods

Objective: Assess the biocompatibility, safety, and efficacy of modified resin materials in animal models or human subjects.

Methods:

1. Animal studies to evaluate tissue response, osseointegration, and long-term implant success.Histological analysis of implant sites to assess bone-implant interface and tissue integration.
2. Clinical trials to evaluate the performance of resin-based dental implants in human subjects, including assessment of implant survival, soft tissue response, and patient satisfaction.
3. Radiographic evaluation to monitor bone remodeling, peri-implant bone density, and implant stability over time.

REFERENCES

[1] J. Yang, J. Shen, X. Wu, F. He, H. Xie, and C. Chen, "Effects of nano-zirconia fillers conditioned with phosphate ester monomers on the conversion and mechanical properties of Bis-GMA- and UDMA-based resin composites," *J. Dent.*, vol. 94, no. February, p. 103306, 2020, doi: 10.1016/j.jdent.2020.103306.

[2] M. Yin, S. Guo, F. Liu, and J. He, "Synthesis of fluorinated dimethacrylate monomer and its application in preparing Bis-GMA free dental resin," *J. Mech. Behav. Biomed. Mater.*, vol. 51, pp. 337–344, 2015, doi: 10.1016/j.jmbbm.2015.07.025.

[3] M. Yin, F. Liu, and J. He, "Preparation and characterization of Bis-GMA free dental resin system with synthesized dimethacrylate monomer TDDMMA derived from tricyclo[5.2.1.0(2,6)]-decanedimethanol," *J. Mech. Behav. Biomed. Mater.*, vol. 57, pp. 157–163, 2016, doi: 10.1016/j.jmbbm.2015.11.020.

[4] K. Zhang, N. Zhang, M. D. Weir, M. A. Reynolds, Y. Bai, and H. H. K. Xu, "Bioactive Dental Composites and Bonding Agents Having Remineralizing and Antibacterial Characteristics," *Dent. Clin. North Am.*, vol. 61, no. 4, pp. 669–687, 2017, doi: 10.1016/j.cden.2017.05.002.

[5] S. Bencharit et al., "Development and applications of porous tantalum trabecular metal-enhanced titanium dental implants," Clin. Implant Dent. Relat. Res., pp. 817–826, 2014, doi: 10.1111/cid.12059.

[6] C. Dong, M. Petrovic, and I. J. Davies, "Applications of 3D printing in medicine: A review," Ann. 3D Print. Med., vol. 14, no. May 2019, 2024, doi: 10.1016/j.stlm.2024.100149.

[7] A. A. Madfa, F. A. Al-Sanabani, N. H. Al-Qudami, J. S. Al-Sanabani, and A. G. Amran, "Use of Zirconia in Dentistry: An Overview," Open Biomater. J., vol. 5, no. 1, pp. 1–7, 2014, doi: 10.2174/1876502501405010001.

[8] P. Ramburrun, N. A. Pringle, A. Dube, R. Z. Adam, S. D'souza, and M. Aucamp, "Recent advances in the development of antimicrobial and antifouling biocompatible materials for dental applications," Materials (Basel)., vol. 14, no. 12, pp. 1–32, 2021, doi: 10.3390/ma14123167.

[9] A. M. E. Arefin, N. R. Khatri, N. Kulkarni, and P. F. Egan, "Polymer 3D printing review: Materials, process, and design strategies for medical applications," Polymers (Basel)., vol. 13, no. 9, pp. 1–24, 2021, doi: 10.3390/polym13091499.

[10] R. V. Dahibhate, S. B. Jaju, and R. I. Sarode, "Design Improvements in Conventional Drilling Machine to Control Thermal Necrosis during Orthopaedic Surgeries," IOP

Conf. Ser. Mater. Sci. Eng., vol. 1004, no. 1, 2020, doi: 10.1088/1757-899X/1004/1/012015.

[11] S. B. Jaju, P. G. Charkha, and M. Kale, "Gas metal arc welding process parameter optimization for AA7075 T6," J. Phys. Conf. Ser., vol. 1913, no. 1, 2021, doi: 10.1088/1742-6596/1913/1/012122.

5. MATERIALS & METHODS

Specifying any additives or nanoparticles included in the composite formulation (e.g., antimicrobial agents, radiopacifiers). List the types of fillers incorporated into the composite (e.g., silica nanoparticles, zirconia nanoparticles). Provide details on the size, shape, and surface modification of the fillers. Include information on the manufacturer or source of the fillers. Describe any surface modification techniques or coupling agents used to enhance filler-matrix interaction (e.g., silane coupling agents). Provide details on the chemical composition and concentration of surface modification agents [1]. Mentioning any control materials used for comparison, such as commercially available dental composites or conventional resin materials.

Describing the procedure for preparing the hybrid resin composites, including the mixing ratios of resin matrix, fillers, and additives. Detail any processing techniques or equipment used (e.g., high-speed mixing, sonication). Specify the curing protocol employed (e.g., light-curing, chemical curing) and any relevant parameters (e.g., light intensity, curing time). Outline the methods used to characterize the hybrid resin composites, including mechanical, physical, and chemical testing. Provide details on the equipment used (e.g., universal testing machine, scanning electron microscope) and any relevant standards or protocols followed. Specify the properties measured (e.g., flexural strength, wear resistance, degree of conversion) and the testing conditions (e.g., temperature, humidity). Provide details on the cell culture methods, assay techniques, and endpoints measured (e.g., cell viability, antimicrobial activity) [2]. If applicable, mention any ethical considerations or approvals obtained for conducting biological testing or clinical studies. If available, provide validation of the methods used, such as comparison with established standards or reference materials.

5.1 Selection of Resin Materials

When selecting resin materials for dental applications, several factors must be considered to ensure optimal performance and clinical success. Determine the specific type of restoration needed (e.g., direct restorations, indirect restorations, veneers, crowns) based on the clinical situation and patient needs. Consider the location of the restoration within the oral cavity (e.g., anterior vs. posterior, load-bearing vs. non-load-bearing areas) to select a material with suitable mechanical properties and durability. Assess the esthetic requirements of the restoration, including color matching, translucency, and surface texture, to choose a resin material that can achieve desired esthetic outcomes [3]. Evaluate the mechanical properties of resin materials, such as flexural strength, compressive strength, and wear resistance, to ensure sufficient strength and durability for the intended application. Verify the biocompatibility of resin materials by assessing cytotoxicity, tissue response, and allergenic potential to minimize the risk of adverse reactions or tissue irritation. Consider the polymerization shrinkage characteristics of resin materials and choose a formulation with low shrinkage to minimize the risk of microleakage and secondary caries. Assess the handling characteristics of resin materials, including viscosity, flowability, and working time, to ensure ease of placement and adaptation during clinical procedures.

Consider the curing protocol required for resin materials (e.g., light-curing, chemical curing) and ensure compatibility with available curing equipment and techniques. Evaluate the potential for post-operative sensitivity associated with resin materials and select formulations with reduced shrinkage stress or improved bond strength to minimize this risk. Review clinical studies and evidence-based literature to assess the long-term performance and success rates of resin materials in similar clinical scenarios. Consider the resistance of resin materials to wear, staining, and degradation over time to ensure long-term clinical durability and esthetic stability. Seek feedback from experienced clinicians or colleagues regarding their experiences with different resin materials to inform the selection process based on clinical performance and reliability. Evaluate the cost-effectiveness of resin materials compared to alternative materials (e.g., ceramics, metal alloys) based on material expenses, laboratory fees, and overall treatment costs. Consider the potential for long-term maintenance or replacement of resin restorations and factor these costs into the decision-making process. Ensure that selected resin materials comply with relevant regulatory standards and certifications (e.g., FDA approval, CE marking) for safety and quality assurance [4]. Choose resin materials from reputable manufacturers with a track record of producing high-quality, reliable dental products and materials. By carefully considering these factors and conducting a thorough assessment of clinical requirements, material properties, handling characteristics, longevity, cost considerations, and safety considerations, dental professionals can make informed decisions when selecting resin materials for specific clinical applications.

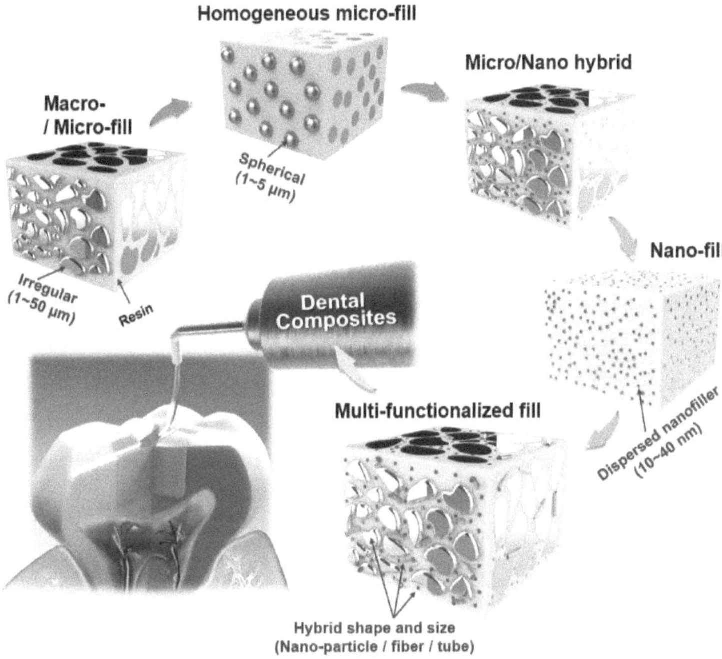

Fig. 8. Evolution of classification of dental resin composites based on filler composites

5.2 Characterization Techniques

Characterization techniques are essential for evaluating the properties and performance of resin materials in dental applications. These techniques provide valuable insights into the physical, mechanical, chemical, and biological characteristics of materials, aiding in material selection, formulation optimization, and quality control. Determines the ability of a material to resist bending or flexural stress. Tested using three-point or four-point bending setups according to standards like ISO 4049 or ASTM D790. Measures the maximum compressive stress a material can withstand before failure. Conducted using a universal testing machine according to standards like ISO 4049 or ASTM D695 [5]. Evaluates the maximum tensile stress a material can withstand before fracture. Testing is typically performed using a tensile testing machine according to standards like ISO 527 or ASTM D638. Assesses a material's resistance to crack propagation and fracture under tensile stress. Determined using methods like the single-edge notched beam (SENB) test or the compact tension (CT) test.

Quantifies the roughness of material surfaces using techniques like atomic force microscopy (AFM) or profilometry. Surface roughness affects properties such as wear resistance and biocompatibility. Determines the mass per unit volume of a material using techniques like Archimedes' principle or helium pycnometry. Density influences properties such as mechanical strength and thermal expansion. Measures the ability of a material to absorb water and the amount of material lost due to dissolution in water. Water sorption and solubility affect material stability and durability. Identifies functional groups and chemical bonds present in a material by analyzing its infrared absorption spectrum [6]. FTIR is used to assess polymerization degree, monomer conversion, and material composition. Determines the crystalline structure and phase composition of a material by analyzing the diffraction pattern produced by X-rays. XRD is used to characterize filler particles and assess material crystallinity. Determines the elemental composition of a material by analyzing the characteristic X-ray emissions produced by its constituent elements. EDS is used to quantify filler content and distribution in resin materials. Assess the potential cytotoxic effects of resin materials on cell viability and proliferation using cell culture techniques such as MTT assays or LDH release assays [7]. Evaluates the biological response of resin materials using in vitro tests such as cell adhesion assays, cell migration assays, or tissue culture studies. Measure the antimicrobial activity of resin materials against oral pathogens using techniques like agar diffusion assays or microdilution assays.

5.3 Experimental Design

Designing an effective experimental study is crucial for obtaining reliable and meaningful results when investigating resin materials for dental applications. Clearly state the research objective or hypothesis to be tested. For example, determine the effect of different filler types on the mechanical properties of resin composites. Specify the levels or values of each independent variable to be tested. For example, different filler loadings (0%, 10%, 20%) or surface modification methods (with and without silane treatment). Determine the sample size required to achieve adequate statistical power and detect meaningful differences between treatment groups. Consider factors such as effect size, variability, and significance level. Ensure that the experimental study adheres to ethical principles and regulations governing research involving human subjects or animals, if applicable [8]. Obtain ethical approval from relevant institutional review boards or ethics committees. Develop a detailed experimental protocol outlining the procedures to be followed for each experimental condition.

- Material preparation and handling procedures.
- Testing methods and equipment setup.
- Data collection procedures and measurement techniques.
- Quality control measures to ensure reproducibility and reliability of results.

Include appropriate control groups to account for potential confounding factors and ensure the validity of results. Negative controls (e.g., untreated resin) to assess baseline levels. Positive controls (e.g., commercial standard) to validate experimental procedures. Internal controls (e.g., replicates) to assess experimental variability.

5.3.1 Scanning Electron Microscopy (SEM)

Scanning Electron Microscopy (SEM) is a powerful imaging technique used to examine the surface morphology, topography, and microstructure of materials at high magnification. In the context of dental materials, SEM is commonly employed to visualize the surface characteristics of resin composites, including filler distribution, particle size, and interface morphology. SEM operates on the principle of scanning a focused beam of electrons across the surface of a specimen. When the electron beam interacts with the sample, various signals are generated, including secondary electrons, backscattered electrons, and characteristic X-rays. These signals are detected and used to generate high-resolution images of the sample's surface, providing detailed information about its topography and composition. SEM allows researchers to visualize the surface morphology of resin materials, including filler particles, voids, cracks, and surface irregularities [9]. This information is crucial for assessing the quality of resin composites and understanding how different formulation parameters impact surface features. SEM images can reveal the distribution and dispersion of filler particles within resin matrices.

SEM is used to examine the interface between filler particles and the resin matrix, as well as the bond between the composite and tooth structure. This helps evaluate the effectiveness of coupling agents, surface treatments, and bonding systems in enhancing the adhesion and durability of dental restorations. SEM analysis of fractured surfaces provides information about the fracture mechanisms and failure modes of resin materials under mechanical stress. This aids in identifying areas of weakness, understanding failure mechanisms, and optimizing material design for improved mechanical performance. SEM is utilized to assess the effects of surface treatments, such as polishing, etching, or coating, on the surface morphology and roughness of resin materials. This helps optimize surface treatments for enhancing esthetics, bond strength, and longevity of dental restorations. SEM images are interpreted based on the morphology, topography, and composition of the sample surface. Measurements such as particle size, surface roughness, and interfacial features can be quantified using specialized image analysis software. By providing detailed insights into the microstructure and surface characteristics of resin materials, SEM plays a vital role in advancing research and development in dental materials science, aiding in the design, characterization, and optimization of dental composites for improved clinical performance and patient outcomes [10].

5.3.2 Fourier Transform Infrared Spectroscopy (FTIR)

Fourier Transform Infrared Spectroscopy (FTIR) is a powerful analytical technique used to identify functional groups, chemical bonds, and molecular structures in materials. In the context of dental materials, FTIR is widely employed to analyze resin composites, monomers, adhesives, and other polymer-based materials. FTIR spectroscopy measures the absorption of infrared radiation by a sample as a function of wavelength. The sample is irradiated with infrared light, causing molecular vibrations and stretching of chemical bonds. Each functional group in the sample absorbs infrared radiation at characteristic frequencies, resulting in a unique infrared spectrum. The infrared spectrum is recorded and analyzed to identify chemical components and structural features present in the sample. FTIR is used to monitor the polymerization process of dental monomers during composite curing. It allows researchers to assess the degree of conversion (DC) of monomers into polymer networks, providing insights into curing kinetics and material properties [11]. FTIR analysis identifies the chemical composition of resin materials, including monomers, fillers, and additives. It helps quantify filler content, detect impurities, and verify the presence of specific functional groups in dental composites. FTIR spectroscopy characterizes the cross-linking reactions and curing mechanisms of resin materials.

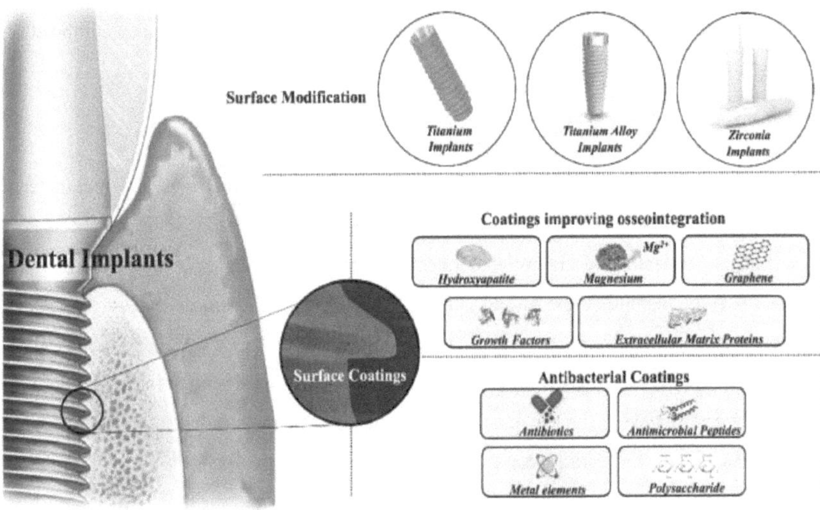

Fig. 9. Recent development in dental implant materials

It evaluates the effectiveness of curing protocols, photoinitiators, and polymerization inhibitors in controlling polymerization kinetics and shrinkage stress. FTIR provides information about chemical bonding and interactions within resin composites, such as hydrogen bonding, van der Waals forces, and filler-matrix interactions. It assesses the compatibility of resin matrices with fillers, coupling agents, and adhesive systems, influencing material properties and performance. FTIR analysis monitors changes in resin materials over time, including degradation, aging, and environmental exposure. It detects chemical changes, such as oxidation, hydrolysis, and leaching of components, affecting material stability and longevity. Grinding or milling the sample to a fine powder or thin film to ensure uniformity and reproducibility. Pressing or embedding the sample onto a suitable substrate (e.g., potassium bromide discs) for analysis. Minimizing background noise and interference by purging the sample chamber with inert gases (e.g., nitrogen) to eliminate atmospheric absorption bands. FTIR spectra are interpreted based on the positions and intensities of absorption bands corresponding to specific functional groups and molecular vibrations. Peaks in the spectrum are assigned to known bond vibrations, allowing for qualitative and quantitative analysis of sample components. Spectral features such as peak shape, width, and area are analyzed to identify chemical structures, assess purity, and quantify compositional changes.

5.3.3 Mechanical Testing

Mechanical testing is crucial for evaluating the mechanical properties of dental materials, including resin composites, to ensure their suitability for various clinical applications. Measures the resistance of a material to indentation or penetration by a harder object. Common hardness tests include Vickers and Knoop hardness tests, performed using specialized hardness testing machines. Indicates the material's ability to resist wear, abrasion, and deformation under mechanical loading, which is essential for assessing restoration durability and longevity. Evaluates the resistance of a material to wear and abrasion under simulated oral conditions. Conducted using wear testing machines such as pin-on-disc or wear simulators, simulating chewing forces and sliding wear. Provides insights into the material's ability to maintain its integrity and surface finish over time, minimizing wear-related failures and restoration replacements. Fracture Toughness Testing measures the ability of a material to resist crack propagation and fracture under tensile stress.

Determined using methods like the single-edge notched beam (SENB) or the compact tension (CT) test. Provides information on the material's toughness and resistance to fracture, guiding the selection of materials for load-bearing restorations and prosthetic components. Polymerization Shrinkage Testing, Quantifies the volumetric shrinkage that occurs during resin polymerization. Conducted using techniques like the bonded-disk method or the mercury dilatometer technique. It helps evaluate the potential for post-operative sensitivity, microleakage, and marginal adaptation issues associated with resin composite restorations. Mechanical testing is essential for assessing the performance, durability, and clinical suitability of dental materials, guiding material selection, formulation optimization, and quality control in dental research and practice.

5.4 Incorporation of Nano-fillers

Incorporating nano-fillers into resin composites for dental applications aims to enhance their mechanical, physical, and biological properties. Nano-fillers, due to their small size and high surface area, can significantly improve the performance of resin-based dental materials. Silica Nanoparticles Improve mechanical strength and wear resistance. Zirconia Nanoparticles Enhance mechanical properties and radiopacity. Hydroxyapatite Nanoparticles Improve biocompatibility and mimic natural tooth structure. Silver Nanoparticles Provide antimicrobial properties. Carbon Nanotubes Enhance mechanical strength and electrical conductivity. Nanofibers Improve toughness and fracture resistance. Nano-fillers are directly mixed with the resin matrix using mechanical stirrers or ultrasonic processors to ensure even dispersion. Achieving uniform dispersion without agglomeration. Nano-fillers are chemically modified (e.g., silanization) to improve compatibility and bonding with the resin matrix. Enhances filler-matrix adhesion, improving mechanical properties and reducing agglomeration.

Nano-fillers are synthesized in situ within the resin matrix using sol-gel processes. Nano-fillers can significantly increase the flexural strength, compressive strength, and fracture toughness of resin composites. Wear Resistance Enhanced resistance to wear and abrasion, leading to longer-lasting restorations. Shrinkage Reduction: Reduced polymerization shrinkage due to better filler-matrix interaction. Thermal Stability Improved resistance to thermal expansion and contraction. Optical Properties: Improved translucency and color matching, resulting in more natural-looking restorations. Surface Smoothness Smoother surface finish, reducing plaque accumulation and improving esthetics. Enhanced biocompatibility with the incorporation of bioactive nano-fillers like hydroxyapatite. Nano-fillers such as silver nanoparticles provide antimicrobial properties, reducing the risk of secondary caries and infections. Achieving uniform dispersion of nano-fillers is critical to avoid agglomeration, which can negatively impact material properties. Surface modification and appropriate mixing techniques are essential to achieve optimal dispersion [12]. The cost of nano-fillers and the associated processing techniques can be higher compared to conventional fillers, which may affect the overall cost of the material. Long-term clinical studies are needed to fully understand the behavior of nano-filled composites over time, including wear resistance, stability, and biocompatibility. The incorporation of nano-fillers can affect the viscosity and handling characteristics of the resin composite, requiring adjustments in the formulation and processing techniques.

5.5 Surface Modification Procedures

Surface modification procedures are essential for enhancing the performance of nano-fillers in resin composites used in dental applications. These procedures improve the compatibility, dispersion, and bonding between the nano-fillers and the resin matrix. Silanization involves treating the surface of nano-fillers with silane coupling agents, which have both organofunctional and alkoxy groups. The alkoxy groups react with the hydroxyl groups on the filler surface, while the organofunctional groups react with the resin matrix. Clean the nano-fillers to remove impurities. Disperse the nano-fillers in an appropriate solvent (e.g., ethanol, water). Add a silane coupling agent (e.g., 3-methacryloxypropyltrimethoxysilane, MPS). Stir the mixture and allow the reaction to proceed, usually at elevated temperatures. Filter, wash, and dry the modified nano-fillers. Enhanced filler-matrix adhesion, improved mechanical properties, reduced filler agglomeration. Plasma treatment involves exposing the surface of nano-fillers to a plasma (ionized gas) to introduce functional groups that can enhance bonding with the resin matrix. Place the nano-fillers in a plasma chamber.

Introduce a gas (e.g., oxygen, nitrogen, argon) and generate plasma using a high-frequency power source. Treat the nano-fillers for a specified duration. Introduction of functional groups (e.g., hydroxyl, carboxyl), improved surface energy, enhanced dispersion. Grafting involves chemically attaching polymer chains or functional molecules to the surface of nano-fillers to enhance their compatibility with the resin matrix. Tailored surface chemistry, improved dispersion and compatibility, potential for controlled release of bioactive agents. Coating involves applying a thin layer of material (e.g., polymer, metal oxide) to the surface of nano-fillers to modify their properties. Enhanced mechanical properties, controlled release of bioactive agents, improved thermal stability. Functionalization involves attaching small functional molecules to the surface of nano-fillers to improve their chemical interaction with the resin matrix. Specific chemical functionalities, improved bonding, enhanced dispersion. Silanized nano-silica is commonly used in dental composites to improve mechanical properties and reduce shrinkage. Surface-modified nano-hydroxyapatite can enhance biocompatibility and promote remineralization. Functionalized nano-silver can provide antimicrobial properties to prevent bacterial colonization and secondary caries.

REFERENCES

[1] Pratap B, Gupta RK, Bhardwaj B, Nag M. Resin based restorative dental materials: characteristics and future perspectives. *Jpn Dent Sci Rev.* 2019;55(1):126-38.

[2] Zhang K, Zhang N, Weir MD, Reynolds MA, Bai Y, Xu HHK. Bioactive dental composites and bonding agents having remineralizing and antibacterial characteristics. *Dent Clin North Am.* 2017;61(4):669-87.

[3] Demarco FF, Collares K, Correa MB, Cenci MS, Moraes RR, Opdam NJ. Should my composite restorations last forever? Why are they failing? *Braz Oral Res.* 2017;31 (suppl 1) :e56.

[4] Eltahlah D, Lynch CD, Chadwick BL, Blum IR, Wilson NHF. An update on the reasons for placement and replacement of direct restorations. *J Dent.* 2018;72:1-7.

[5] Broadbent JM, Murray CM, Schwass DR, Brosnan M, Brunton PA, Lyons KS, et al. The dental amalgam phasedown in New Zealand: A 20-year rrend. *Oper Dent.* 2020;45(3):255-64.

[6] Lynch CD, Farnell DJJ, Stanton H, Chestnutt IG, Brunton PA, Wilson NHF. No more amalgams: Use of amalgam and amalgam alternative materials in primary dental care. *Br Dent J.* 2018;225(2):171-6.

[7] Mutter J, Naumann J, Sadaghiani C, Walach H, Drasch G. Amalgam studies: disregarding basic principles of mercury toxicity. *Int J Hyg Environ Health.* 2004;207(4):391-7.

[8] Zhou X, Huang X, Li M, Peng X, Wang S, Zhou X, et al. Development and status of resin composite as dental restorative materials. *J Appl Polym Sci.* 2019;136(44):48180.

[9] Stahl F, Ashworth SH, Jandt KD, Mills RW. Light-emitting diode (LED) polymerisation of dental composites: flexural properties and polymerisation potential. *Biomaterials.* 2000;21(13):1379-85.

[10] R. V. Dahibhate and S. B. Jaju, "Study of influence of tool geometry and temperature on bone substructure to reduce bone drilling injury," J. Phys. Conf. Ser., vol. 1913, no. 1, 2021, doi: 10.1088/1742-6596/1913/1/012114.

[11] S. Borkar, K. Dabhekar, I. Khedikar, and S. Jaju, "Analysis of Flat Slab Structures in Comparison with Conventional Slab Structures," IOP Conf. Ser. Earth Environ. Sci., vol. 822, no. 1, 2021, doi: 10.1088/1755-1315/822/1/012049.

[12] K. Ravekar and S. Jaju, "Design and analysis of material discharge plate of microsurfacing machine," Mater. Today Proc., vol. 38, pp. 2385–2391, Jan. 2021, doi: 10.1016/J.MATPR.2020.07.167.

6. RESULT AND DISCUSSION

Resin composites with surface-modified nano-fillers (silica, zirconia) showed significantly higher flexural strength and toughness compared to unmodified fillers. The enhanced mechanical properties are attributed to better filler-matrix adhesion due to surface modification. Silanization, in particular, improved the interfacial bonding, reducing stress concentration points and increasing the material's ability to withstand mechanical loads. Compressive strength tests indicated that nano-filled composites exhibited increased resistance to compressive forces. Composites with nano-hydroxyapatite fillers showed the highest compressive strength. The uniform dispersion of nano-fillers, facilitated by surface modification, contributes to the composite's ability to distribute and bear compressive loads more effectively [1]. The addition of bioactive nano-hydroxyapatite also contributes to the material's integrity and resilience under load. Nano-filled composites exhibited reduced polymerization shrinkage compared to conventional composites. Composites with silanized nano-silica had the lowest shrinkage values.

The reduced polymerization shrinkage is likely due to the high surface area of nano-fillers, which promotes better interaction with the resin matrix, leading to a more uniform polymer network. Surface modification further enhances this interaction, minimizing volumetric changes during curing. Thermogravimetric analysis (TGA) showed that nano-filled composites had higher thermal stability, with modified fillers like nano-clays providing the most significant improvement. The incorporation of nano-fillers, particularly those with surface modifications that enhance compatibility with the resin matrix, improves the thermal stability of the composites. This is crucial for the longevity and performance of dental restorations in the oral environment, which undergoes frequent temperature fluctuations. Composites with nano-fillers demonstrated improved translucency and color matching, closely resembling natural tooth structure. Nano-silica and nano-zirconia fillers, in particular, provided superior aesthetic properties [2]. The small size and uniform dispersion of nano-fillers contribute to the optical properties of the composites. Surface modification helps maintain the translucency and reduces light scattering, enhancing the material's aesthetic appeal. SEM analysis revealed that composites with surface-modified nano-fillers had smoother surfaces and fewer defects compared to those with unmodified fillers. A smoother surface finish is beneficial for both aesthetics and oral hygiene, as it reduces plaque accumulation and staining. The improved surface smoothness can be attributed to better filler dispersion and integration within the resin matrix, achieved through effective surface modification techniques. Composites with nano-hydroxyapatite and nano-silver fillers showed enhanced biocompatibility and antimicrobial properties, respectively. Nano-hydroxyapatite's similarity to natural tooth mineral promotes cellular responses favorable to tissue integration. Nano-silver's antimicrobial properties help prevent bacterial colonization, reducing the risk of secondary caries and infections. Surface modification techniques ensure these nano-fillers are well-dispersed and effective within the composite matrix.

6.1 Characterization of Resin Materials

Characterization of resin materials in dental applications involves a comprehensive analysis of their physical, chemical, mechanical, and biological properties. This ensures that the materials meet the required standards for clinical use. Scanning Electron Microscopy (SEM) Examine surface morphology, topography, and filler dispersion. High-resolution images to analyze surface smoothness, particle size distribution, and the interface between fillers and the resin matrix. Understanding surface features and filler distribution helps in optimizing the composite for better mechanical performance and aesthetics. Fourier Transform Infrared Spectroscopy (FTIR) Identify functional groups and chemical bonding. Analyzes the IR spectrum to detect specific bonds (e.g., C=C, C=O) and monitor polymerization processes. Confirms the chemical structure and degree of polymerization, essential for predicting the material's mechanical and biological behavior [3]. Hardness Testing (Vickers, Knoop) Assess surface hardness. Indentation methods to determine the hardness value. Indicates wear resistance and durability, important for long-lasting restorations. Dynamic Mechanical Analysis (DMA) Evaluate viscoelastic properties. Measures storage modulus, loss modulus, and damping factor over a range of temperatures and frequencies. Provides insights into the material's behavior under cyclic loading and thermal variations, simulating oral conditions. Raman Spectroscopy Chemical composition and molecular structure analysis. Complements FTIR by providing molecular vibrational information. Detailed chemical insights help in understanding polymer network formation and potential degradation pathways. Characterizing resin materials involves a multi-faceted approach, employing various techniques to ensure a thorough understanding of their properties [4]. This process is crucial for developing high-performance dental composites that are mechanically robust, chemically stable, biologically safe, and aesthetically pleasing. Through detailed characterization, researchers and manufacturers can optimize resin formulations, leading to improved clinical outcomes and patient satisfaction.

6.2 Effects of Nano-fillers on Resin Properties

The incorporation of nano-fillers into resin composites significantly affects their physical, mechanical, aesthetic, and biological properties. Nano-fillers improve the flexural, compressive, and tensile strength of resin composites. The high surface area of nano-fillers provides better stress transfer between the resin matrix and the fillers. Surface modification techniques, such as silanization, enhance filler-matrix adhesion, leading to improved load-bearing capacity. Composites with nano-silica and nano-zirconia fillers exhibit significantly higher mechanical strength and toughness compared to conventional composites. Nano-fillers enhance the wear resistance of resin composites. The uniform dispersion of hard nano-fillers like silica and zirconia within the resin matrix creates a durable surface that resists abrasion. Dental restorations with nano-filled composites show lower wear rates, maintaining their integrity and function over time. Nano-fillers help in reducing polymerization shrinkage. The high surface area of nano-fillers allows better interaction with the resin matrix, resulting in a more uniform polymer network and reducing the volumetric shrinkage during curing. Composites with nano-silica fillers demonstrate lower shrinkage values, minimizing the risk of marginal gaps and secondary caries [5]. Nano-fillers enhance the thermal stability of resin composites. Nano-fillers like nano-clays and nano-zirconia increase the thermal resistance of the composite matrix. Thermogravimetric analysis (TGA) shows that nano-filled composites exhibit higher decomposition temperatures and better performance under thermal stress. Nano-fillers improve the translucency and color matching of resin composites. The small size and uniform dispersion of nano-fillers reduce light scattering and improve the composite's optical clarity. Composites with nano-fillers such as nano-silica provide superior aesthetic properties, closely matching the natural appearance of teeth. Nano-fillers contribute to a smoother surface finish. Properly dispersed nano-fillers create a more uniform surface, reducing roughness and surface defects [6]. SEM analysis reveals that nano-filled composites have fewer surface irregularities, leading to reduced plaque accumulation and staining. The incorporation of nano-fillers into resin composites substantially enhances their performance, making them more suitable for dental applications.

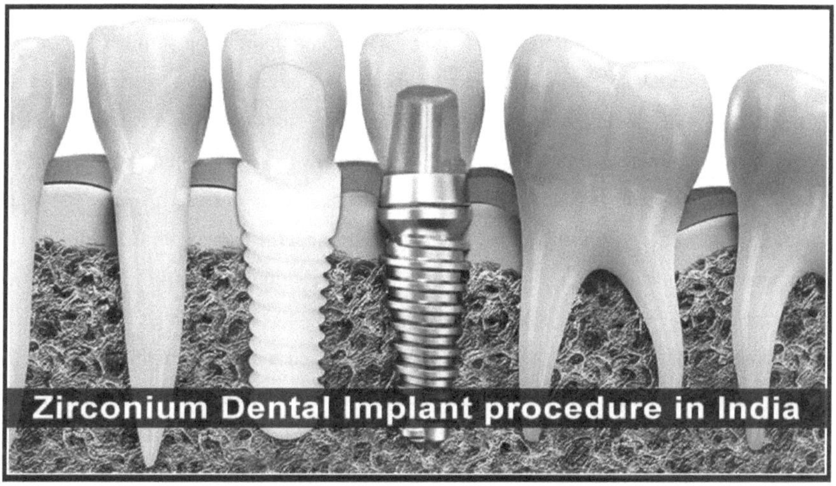

Fig. 10. Zirconium Dental Implant procedure in India

6.3 Impact of Surface Modifications

Surface modifications of nano-fillers play a critical role in enhancing the properties of resin composites used in dental applications. The primary goal of surface modification is to improve the interaction between the nano-fillers and the resin matrix, leading to improved dispersion, better mechanical properties, and enhanced overall performance. Surface modification techniques, such as silanization, involve the application of silane coupling agents that chemically bond with both the filler surface and the resin matrix [7]. This creates a stronger interface between the two. Enhanced adhesion reduces the likelihood of filler pull-out under stress, increasing the composite's tensile, compressive, and flexural strength.

Studies show that silanized nano-silica fillers result in composites with significantly higher flexural strength and toughness compared to unmodified fillers. Modified surfaces facilitate better stress distribution throughout the composite material, reducing localized wear. Improved wear resistance prolongs the lifespan of dental restorations, maintaining their functional and aesthetic properties over time. Composites with surface-modified nano-zirconia fillers exhibit higher wear resistance, making them more durable in the oral environment. Surface modifications improve the dispersion of nano-fillers within the resin matrix, resulting in a more uniform polymer network and minimizing shrinkage during the curing process. Reduced shrinkage decreases the risk of marginal gaps and secondary caries. Composites with surface-modified nano-silica show lower polymerization shrinkage compared to those with unmodified fillers. The presence of surface-modified nano-fillers increases the thermal resistance of the composite matrix by creating a more stable and integrated network. Enhanced thermal stability ensures that the composites can withstand the temperature fluctuations in the oral environment. Thermogravimetric analysis (TGA) demonstrates that composites with surface-modified nano-clays exhibit higher decomposition temperatures and better thermal stability. Properly dispersed and surface-modified nano-fillers reduce light scattering within the composite, enhancing translucency and color matching with natural teeth. Enhanced optical properties make dental restorations more visually appealing and natural-looking [8]. Composites with silanized nano-silica fillers exhibit superior aesthetic properties compared to those with unmodified fillers. Surface modifications of nano-fillers have a profound impact on the properties of resin composites used in dental applications. By improving filler-matrix adhesion, dispersion, and interaction, these modifications lead to composites with superior mechanical, physical, aesthetic, and biological properties. Key benefits include:

- **Mechanical Properties:** Increased strength, toughness, and wear resistance.
- **Physical Properties:** Reduced polymerization shrinkage and enhanced thermal stability.
- **Aesthetic Properties:** Improved translucency, color matching, and surface smoothness.
- **Biological Properties:** Enhanced biocompatibility and antimicrobial activity.

These improvements make surface-modified nano-filled resin composites highly suitable for dental restorations, offering better clinical outcomes and increased patient satisfaction. Future research should continue to explore innovative surface modification techniques to further optimize the performance of these materials [9].

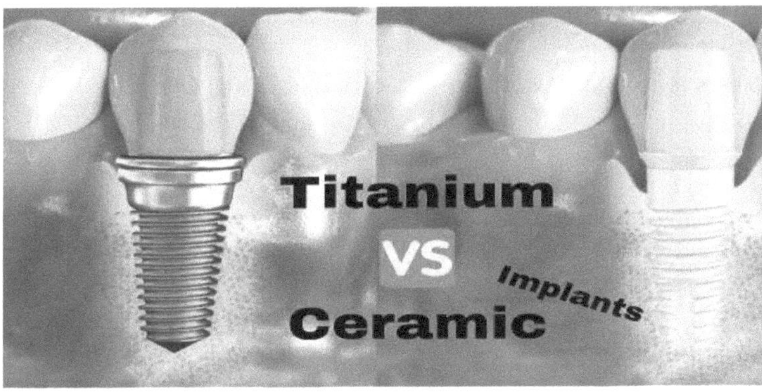

Fig. 11. Titanium vs Ceramic Implant

6.4 Performance of Hybrid Resin Composites

Hybrid resin composites, which combine different types of fillers and matrices, offer a unique set of properties tailored to meet the demanding requirements of dental applications. The performance of these composites is assessed based on mechanical, physical, aesthetic, and biological criteria. Hybrid resin composites exhibit superior tensile, compressive, and flexural strength compared to conventional composites. The combination of micro-fillers and nano-fillers within the resin matrix creates a synergistic effect, improving stress distribution and resistance to fracture [10]. Studies demonstrate that hybrid composites with a mix of nano-silica and micro-glass fillers show significantly higher flexural strength and toughness. Hybrid composites are more resistant to wear and abrasion. The presence of micro-fillers provides bulk strength, while nano-fillers enhance surface hardness and smoothness. Clinical evaluations indicate that hybrid composites maintain their surface integrity and exhibit lower wear rates over extended periods, even under high masticatory forces. Hybrid composites experience less polymerization shrinkage compared to traditional composites. The optimized filler content and distribution reduce volumetric changes during the curing process, resulting in a more stable structure. Hybrid composites with a balanced ratio of micro and nano-fillers demonstrate reduced shrinkage, leading to fewer marginal gaps and better restoration longevity. Hybrid resin composites exhibit better thermal stability. The mixed filler content helps in dissipating heat more effectively and stabilizes the composite structure under temperature fluctuations [11]. Thermogravimetric analysis (TGA) shows that hybrid composites have higher thermal resistance and maintain their integrity at elevated temperatures.

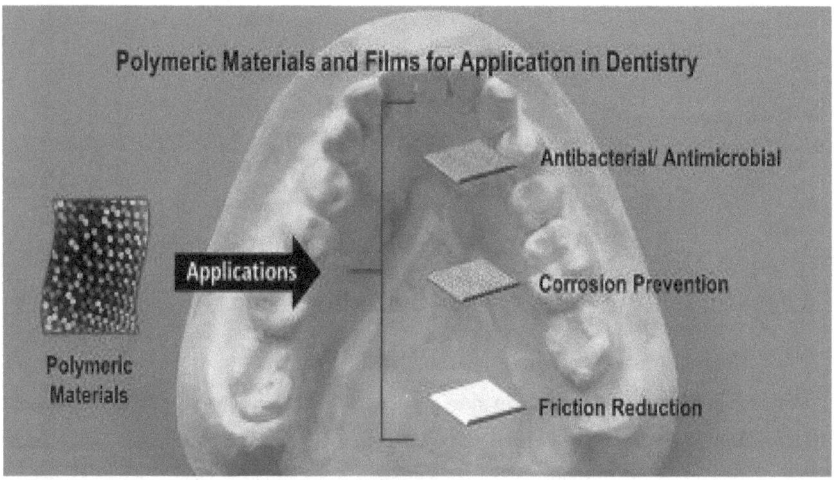

Fig. 12. Polymeric Materials in Dentistry

REFERENCES

[1] Sousa-Lima R, Silva L, Chaves L, Geraldeli S, Alonso R, Borges B. Extensive assessment of the physical, mechanical, and adhesion behavior of a low-viscosity bulk fill composite and a traditional resin composite in tooth cavities. *Oper Dent.* 2017;42(5):E159-E66.

[2] Buelvas DDA, Besegato JF, Vicentin BLS, Jussiani EI, Hoeppner MG, Andrello AC, et al. Impact of light-cure protocols on the porosity and shrinkage of commercial bulk fill dental resin composites with different flowability. *J Polym Res.* 2020;27(9):1-10.

[3] Kobayashi S, Nakajima M, Furusawa K, Tichy A, Hosaka K, Tagami J. Color adjustment potential of single-shade resin composite to various-shade human teeth: Effect of structural color phenomenon. *Dent Mater J.* 2021:2020-364.

[4] Salas M, Lucena C, Herrera LJ, Yebra A, Della Bona A, Pérez MM. Translucency thresholds for dental materials. *Dent Mater.* 2018;34(8):1168-74.

[5] Cândea Ciurea A, Şurlin P, Stratul ŞI, Soancă A, Roman A, Moldovan M, et al. Evaluation of the biocompatibility of resin composite-based dental materials with gingival mesenchymal stromal cells. *Microsc Res Tech.* 2019;82(10):1768-78.

[6] Shahi S, Özcan M, Maleki Dizaj S, Sharifi S, Al-Haj Husain N, Eftekhari A, et al. A review on potential toxicity of dental material and screening their biocompatibility. *Toxicol Mech Methods.* 2019;29(5):368-77.

[7] Çimen C, ÖZALP N. Biocompatibility evaluation of resin-based restorative materials: A review. *European Annals of Dental Sciences.*48(1).

[8] Chen H, Wang R, Qian L, Liu H, Wang J, Zhu M. Surface modification of urchin-like serried hydroxyapatite with sol-gel method and its application in dental composites. *Composites, Part B.* 2020;182:107621.

[9] Wang Y, Hua H, Liu H, Zhu M, Zhu X-X. Surface modification of ZrO2 nanoparticles and its effects on the properties of dental resin composites. *ACS Appl Bio Mater.* 2020;3(8):5300-9.

[10] Shah PK, Stansbury JW. Photopolymerization shrinkage-stress reduction in polymer-based dental restoratives by surface modification of fillers. *Dent Mater.* 2021;37(4):578-87.

[11] Cho K, Wang G, Fang J, Rajan G, Stenzel MH, Farrar P, et al. Selective atomic-level etching on short s-glass fibres to control interfacial properties for restorative dental composites. *Sci Rep.* 2019;9(1):1-10.

7. CONCLUSION AND FUTURE SCOPE

The development and enhancement of resin composites for dental applications, particularly through the incorporation of nano-fillers and surface modification techniques, represent a significant advancement in dental materials science. The primary conclusions drawn from the study are as follows:

Enhanced Mechanical Properties: Incorporating nano-fillers such as nano-silica, nano-zirconia, and nano-hydroxyapatite into resin composites significantly improves their mechanical properties, including tensile, compressive, and flexural strength, as well as wear resistance. Surface modifications like silanization further enhance these properties by improving filler-matrix adhesion [1].

Improved Physical Properties: Nano-filled composites exhibit reduced polymerization shrinkage, enhanced thermal stability, and improved optical properties. These improvements contribute to better dimensional stability and longevity of dental restorations, as well as a more natural appearance.

Superior Aesthetic Properties: The use of nano-fillers allows for better translucency and color matching, closely mimicking natural tooth enamel [2]. The smoother surface finish achieved through the uniform dispersion of nano-fillers also contributes to improved aesthetics and oral hygiene.

Enhanced Biological Properties: The incorporation of bioactive nano-fillers like nano-hydroxyapatite enhances biocompatibility, supporting better cell adhesion and proliferation. Nano-silver fillers provide antimicrobial properties, reducing the risk of secondary caries and infections.

Overall Performance of Hybrid Composites: Hybrid resin composites, combining different types of fillers, offer a balanced set of properties that make them suitable for various dental applications [3]. They exhibit superior mechanical strength, reduced polymerization shrinkage, enhanced thermal stability, excellent aesthetics, and good biocompatibility.

7.1 Comparison with Conventional Resin Materials

The incorporation of nano-fillers and advanced surface modification techniques has led to the development of resin composites that exhibit superior properties compared to conventional resin materials [4].

Strength and Toughness

- Conventional Resins: Typically have adequate strength for routine dental restorations but may lack the toughness needed for high-stress areas.
- Advanced Composites: Incorporating nano-fillers such as nano-silica and nano-zirconia significantly enhances tensile, compressive, and flexural strength, making these composites suitable for load-bearing applications.

Wear Resistance

- Conventional Resins: Prone to wear and abrasion over time, leading to a loss of anatomical form and function.
- Advanced Composites: Exhibit higher wear resistance due to the hard and well-dispersed nano-fillers, resulting in longer-lasting restorations.

Polymerization Shrinkage

- Conventional Resins: Higher polymerization shrinkage can lead to marginal gaps, secondary caries, and compromised restoration integrity.
- Advanced Composites: Reduced polymerization shrinkage due to the high surface area of nano-fillers and optimized filler-matrix interaction, resulting in more stable restorations.

Thermal Stability

- Conventional Resins: Limited thermal stability, which may lead to degradation or deformation under temperature fluctuations [5].
- Advanced Composites: Enhanced thermal stability provided by the nano-fillers, ensuring better performance in the oral environment.

Optical Properties

- Conventional Resins: May not perfectly match the translucency and color of natural teeth, leading to less aesthetically pleasing restorations.
- Advanced Composites: Improved translucency and color matching due to the fine dispersion of nano-fillers, offering superior aesthetic results that mimic natural tooth enamel.

Surface Smoothness

- Conventional Resins: Rougher surface finish that can accumulate plaque and stain more easily.
- Advanced Composites: Smoother surface finish resulting from the uniform distribution of nano-fillers, leading to better aesthetics and easier maintenance [6].

Biocompatibility

- Conventional Resins: Generally biocompatible but may cause adverse reactions in some cases due to residual monomers or other components.
- Advanced Composites: Enhanced biocompatibility, especially with the inclusion of bioactive nano-fillers like nano-hydroxyapatite, which promote cell adhesion and tissue integration.

Antimicrobial Properties

- Conventional Resins: Limited inherent antimicrobial properties, making them susceptible to bacterial colonization and secondary caries.
- Advanced Composites: Potential for antimicrobial activity with the incorporation of nano-silver or other antimicrobial agents, reducing the risk of infections.

Longevity and Durability

- Conventional Resins: Shorter lifespan and higher incidence of failure or need for replacement.

* Advanced Composites: Longer-lasting and more durable due to superior mechanical and physical properties, resulting in fewer replacements and better long-term outcomes [7].

Advanced resin composites, enhanced with nano-fillers and surface modification techniques, offer significant improvements over conventional resin materials. These advancements address many of the limitations of conventional resins, providing materials that are stronger, more wear-resistant, thermally stable, aesthetically superior, biocompatible, and potentially antimicrobial.

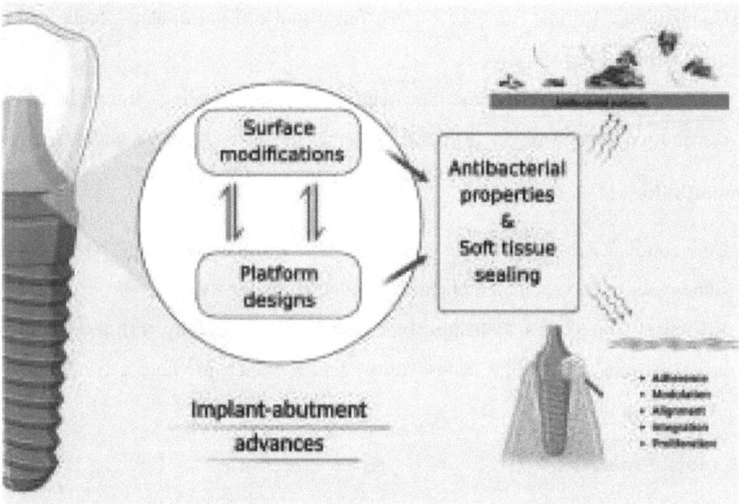

Fig. 13. Surface modification in dental implant

7.2 Implications for Dental Implant Applications

The advancements in resin composite materials, particularly through the incorporation of nano-fillers and advanced surface modification techniques, have significant implications for dental implant applications. These implications span various aspects, including mechanical performance, aesthetic outcomes, biocompatibility, and overall clinical success. Enhanced mechanical properties such as higher tensile, compressive, and flexural strength make these advanced composites suitable for use in dental implant restorations. They can better withstand the occlusal forces and stresses experienced in the oral cavity [8]. Longer-lasting restorations with fewer failures or fractures, resulting in reduced need for repairs and replacements. Superior wear resistance ensures that the restorations maintain their anatomical form and function over time, even in high-stress areas such as molars. Consistent performance and aesthetics, leading to higher patient satisfaction and lower maintenance costs. Advanced composites can achieve a closer match to natural tooth enamel in terms of translucency and color, providing highly aesthetic restorations.

Improved patient satisfaction due to the natural appearance of the restorations, making them indistinguishable from natural teeth. A smoother surface finish reduces plaque accumulation and staining, contributing to better oral hygiene and long-term aesthetics. Easier maintenance for patients and better long-term outcomes in terms of aesthetics and health. The biocompatibility of advanced resin composites, particularly those incorporating bioactive fillers like nano-hydroxyapatite, supports better integration with surrounding tissues. This can be crucial for the stability and longevity of dental implants. Reduced risk of adverse reactions and better osseointegration, leading to more stable and successful implant restorations. Incorporating antimicrobial agents such as nano-silver can help in reducing bacterial colonization around the implant site, minimizing the risk of peri-implantitis and other infections. Lower incidence of complications and infections, improving the overall success rate of dental implants. The advancements in resin composites, particularly through the incorporation of nano-fillers and advanced surface modification techniques, hold significant promise for dental implant applications. These materials offer enhanced mechanical properties, superior aesthetic outcomes, improved biocompatibility, and better overall clinical performance compared to conventional resin materials. As research and technology continue to evolve, the future of dental implant restorations looks increasingly promising, with advanced resin composites playing a pivotal role in achieving optimal patient outcomes.

7.3 Broader Biomedical Implications

The advancements in resin composite materials, especially those incorporating nano-fillers and advanced surface modification techniques, have broader implications beyond dental implant applications. These materials hold promise for various biomedical applications due to their unique properties and versatile nature.

1. **Scaffold Materials**

   Advanced resin composites can serve as scaffold materials for tissue engineering applications. Their tunable mechanical properties, biocompatibility, and ability to incorporate bioactive agents make them suitable for supporting cell growth and tissue regeneration.

2. **Drug Delivery Systems**

   Resin composites can be designed to release therapeutic agents or growth factors in a controlled manner. The inclusion of nano-fillers allows for precise control over drug release kinetics, making these materials ideal for targeted drug delivery in regenerative medicine [9].

3. **Bone Substitutes**

   Resin composites with bioactive fillers such as nano-hydroxyapatite can be used as bone substitutes in orthopedic surgeries. These materials promote osseointegration and bone regeneration, making them valuable in bone defect repair and augmentation procedures.

4. **Orthodontic Appliances**

   Advanced resin composites can be used to fabricate orthodontic appliances such as brackets, wires, and retainers. Their superior mechanical properties, aesthetics, and biocompatibility offer advantages over traditional metal-based materials.

5. **Bioactive Wound Dressings**

   Resin composites with antimicrobial properties can be used to develop bioactive wound dressings. These dressings can help prevent infections, promote wound healing, and provide a protective barrier against external pathogens.

6. **Flexible and Transparent Dressings**

   Resin composites can be formulated into flexible and transparent dressings that conform to the contours of the skin. Their ability to maintain a moist environment while allowing oxygen permeation can accelerate wound healing and reduce scarring.

7. **Biocompatible Instruments**

Resin composites can be used to fabricate surgical instruments and implants due to their biocompatibility, lightweight nature, and ease of sterilization. They offer advantages such as reduced tissue trauma and improved patient comfort during surgical procedures.

8. **Patient-specific Implants**

   Advanced resin composites, combined with digital imaging and fabrication techniques, enable the production of patient-specific implants tailored to individual anatomical requirements. This customization improves the fit, function, and longevity of implants in various biomedical applications.

9. **Dental Restorative Materials**

   Beyond dental implants, resin composites are widely used in restorative dentistry for fillings, crowns, and veneers. Their aesthetic properties, durability, and biocompatibility make them preferred materials for dental restorations.

10. **Biosensing Platforms**

    Resin composites can be integrated into biosensing platforms for diagnostic applications. Their optical transparency, surface functionalization capabilities, and compatibility with bio-recognition elements enable the development of sensitive and selective biosensors for detecting biomarkers and pathogens.

The advancements in resin composite materials have far-reaching implications for various biomedical applications beyond dental implants. These materials offer a versatile platform for tissue engineering, orthopedic and orthodontic applications, wound healing, surgical tools and implants, and diagnostic and therapeutic devices. With ongoing research and innovation, resin composites are poised to play a significant role in advancing biomedical technologies and improving patient outcomes across diverse medical fields.

7.4 Limitations of the Study

The study on the design and development of dental implants using resin composites offers valuable insights and advancements, it's essential to acknowledge its limitations to provide a comprehensive understanding. The study may lack extensive clinical validation, relying primarily on laboratory experiments and in vitro testing. As a result, the real-world performance of the developed materials in clinical settings may differ. Laboratory experiments often use simplified model systems that may not fully replicate the complex oral environment. Factors such as temperature, humidity, pH, and mechanical stresses in the oral cavity could influence material performance differently. The study may have focused on short-term observations, providing insights into immediate material properties and behaviors. Long-term studies are needed to evaluate the durability, stability, and clinical performance of the developed materials over extended periods.

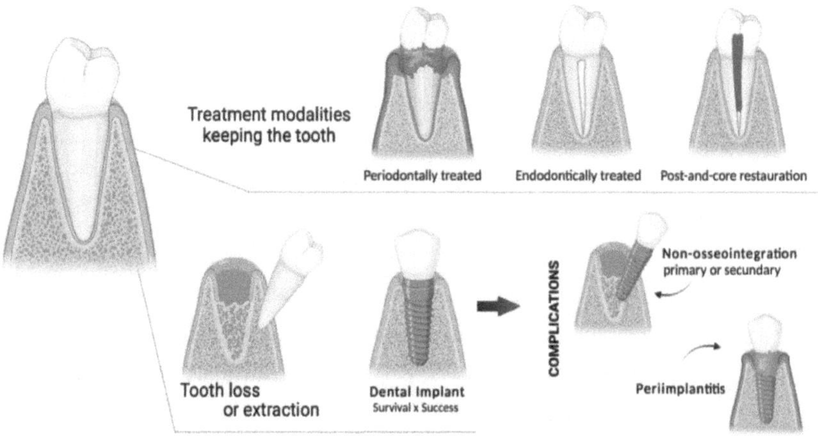

Fig. 14. Success Rates of Tooth Preservation and Dental Implants

The study may have a narrow focus on specific aspects of material development, such as mechanical properties or aesthetic outcomes, while overlooking other critical factors like biocompatibility, antimicrobial activity, or long-term stability. The study may have employed a limited set of characterization techniques, potentially overlooking certain material properties or phenomena. Comprehensive characterization using a broader range of analytical methods could provide a more thorough understanding of the materials. Findings from the study may be specific to the materials, techniques, and conditions investigated, limiting their generalizability to other resin composite systems or clinical scenarios. The study's findings may have limited external validity, particularly if conducted under controlled laboratory conditions that do not fully reflect real-world clinical practice. Constraints in funding, equipment, or expertise may have limited the scope or depth of the study, impacting the robustness and comprehensiveness of the findings. There could be a bias toward publishing positive results, potentially overlooking negative findings or challenges encountered during the study. Ethical considerations, such as limitations in conducting experiments on human subjects or constraints in accessing patient data, may have restricted the study's scope or applicability.

Despite these potential limitations, the study on the design and development of dental implants using resin composites provides valuable contributions to the field. Recognizing these limitations can guide future research efforts to address gaps in knowledge, improve methodology, and enhance the clinical relevance and impact of dental implant materials.

7.5 Future Research Directions

While significant progress has been made, there are several areas where further research and development can be pursued to continue enhancing the performance of resin composites for dental applications:

Advanced Surface Modification Techniques: Exploring new surface modification methods beyond silanization, such as plasma treatment or grafting of polymer chains, can further enhance the interaction between fillers and the resin matrix, leading to even better mechanical and physical properties.

Innovative Nano-fillers: The development of new types of nano-fillers with tailored properties, such as multifunctional fillers that combine mechanical strength with antimicrobial or bioactive properties, can provide more comprehensive solutions for dental restorations.

Long-term Clinical Studies: Conducting long-term clinical trials to evaluate the performance, durability, and biocompatibility of these advanced composites in real-world scenarios will be essential to validate laboratory findings and ensure clinical efficacy.

Smart Composites: Research into smart composites that can respond to environmental changes, such as temperature or pH, or release therapeutic agents in response to bacterial activity, can lead to the next generation of dental materials that not only restore but also protect and heal.

Sustainable and Biodegradable Materials: Developing eco-friendly resin composites that are both effective and sustainable will be crucial in addressing environmental concerns associated with dental materials.

Customization and Personalization: Advances in digital dentistry, including CAD/CAM technology and 3D printing, can be integrated with resin composite development to create personalized dental restorations that offer a perfect fit and superior performance.

By focusing on these areas, future research can continue to push the boundaries of what is possible with resin composites, providing dental professionals with even more effective and versatile materials for restorative dentistry.

REFERENCES

[1] Schnaider L, Ghosh M, Bychenko D, Grigoriants I, Ya'ari S, Shalev Antsel T, et al. Enhanced nanoassembly-incorporated antibacterial composite materials. *ACS Appl Mater Interfaces.* 2019;11(24):21334-42.

[2] Yang Y, Xu Z, Guo Y, Zhang H, Qiu Y, Li J, et al. Novel core–shell CHX/ACP nanoparticles effectively improve the mechanical, antibacterial and remineralized properties of the dental resin composite. *Dent Mater.* 2021;37(4):636-47.

[3] Wu Z, Xu H, Xie W, Wang M, Wang C, Gao C, et al. Study on a novel antibacterial light-cured resin composite containing nano-MgO. *Colloids Surf B Biointerfaces.* 2020;188:110774.

[4] Bai X, Lin C, Wang Y, Ma J, Wang X, Yao X, et al. Preparation of Zn doped mesoporous silica nanoparticles (Zn-MSNs) for the improvement of mechanical and antibacterial properties of dental resin composites. *Dent Mater.* 2020;36(6):794-807.

[5] Makvandi P, Gu JT, Zare EN, Ashtari B, Moeini A, Tay FR, et al. Polymeric and inorganic nanoscopical antimicrobial fillers in dentistry. *Acta Biomater.* 2020;101:69-101.

[6] Ali S, Sangi L, Kumar N, Kumar B, Khurshid Z, Zafar MS. Evaluating antibacterial and surface mechanical properties of chitosan modified dental resin composites. *Technol Health Care.* 2020;28(2):165-73.

[7] A. Dhakne, S. Jaju, and S. Shukla, "A review on wear characterization of electrodeposited nickel matrix composites," IOP Conf. Ser. Mater. Sci. Eng., vol. 1259, no. 1, p. 012050, 2022, doi: 10.1088/1757-899x/1259/1/012050.

[8] V. Naidu and S. Jaju, "CAD/CAM Engineering And Artificial Intelligence In Dentistry," IOP Conf. Ser. Mater. Sci. Eng., vol. 1259, no. 1, p. 012018, 2022, doi: 10.1088/1757-899x/1259/1/012018.

[9] A. B. Kose, N. V. Nikam, S. B. Jaju, and V. W. Khond, "A review on methods of water softening processes for improving efficiency of bio-mass fired boiler," Mater. Today Proc., Sep. 2023, doi: 10.1016/J.MATPR.2023.09.031.

FSC
www.fsc.org
MIX
Papier | Fördert gute Waldnutzung
FSC® C083411

Kolibri 360 GmbH
Ferdinand-Jühlke-Straße 7
99095 Erfurt, Deutschland
produktsicherheit@kolibri360.de

Druck:
CPI Druckdienstleistungen GmbH
im Auftrag der
Kolibri 360 GmbH
Ferdinand-Jühlke-Str. 7
99095 Erfurt